T0300984

"This seminal book represents an excellent resource on the selection of materials for rivets in aerospace applications. Through rigorous modelling, insightful materials science, and compelling case studies, the authors masterfully dissect the pivotal role of rivets in aerospace engineering and offer profound insights into both the triumphs and tragedies of the industry, including the iconic Concorde. This highly readable text is suitable for students, educators, and aerospace engineers."

– **Michael Ferry**, *Professor & Head, School of Materials Science & Engineering, University of New South Wales (UNSW)*

"This book fills a big void. It is focused on a technical subject that is easily overlooked in typical education curricula. The book goes into a level of detail, found nowhere else, covering fundamental and applied aspects. An important aspect is that much of the technical content remains relevant to hypersonic flight, which is relevant in a host of military applications. ... Engineers would find this book an invaluable resource for designing rivets that can withstand the extreme conditions of supersonic flight."

– **Amine Benzerga**, *Professor, Aerospace Engineering and Materials Science & Engineering | General Dynamics Professor in Aerospace Engineering, Texas A&M University*

Supersonic Effects on Rivets

Supersonic Effects on Rivets introduces aerospace components, such as rivets, used in subsonic and supersonic/hypersonic aircraft. It investigates the various alloys used to manufacture rivets/fasteners and the heat treatment of those alloys.

Providing background on commercial (subsonic and supersonic) and military (subsonic and supersonic/hypersonic) aircraft, the book discusses selecting materials, rivet arrangement, skin friction/drag effects, estimating temperature, thermal properties, and fatigue testing of aerospace rivets. It includes real-world case studies on aircraft failures due to incorrect design and failure mechanisms of aerospace rivets. Lessons learnt from the failures of the iconic Concorde, Space Shuttle Columbia, and American Airbus A300-600 Flight 587 are also deliberated upon. Laboratory research including future recommendations are presented.

The book will be useful for military applications, commercial aircraft, practicing aerospace/aeronautical engineers, materials scientists, metallurgical engineers, universities, colleges, higher education, schools and students.

Supersonic Effects on Rivets

George Nadim Melhem, Paul Richard Munroe
and Akshay Vithal

CRC Press
Taylor & Francis Group
Boca Raton London New York

CRC Press is an imprint of the
Taylor & Francis Group, an **informa** business

First edition published 2025
by CRC Press
2385 NW Executive Center Drive, Suite 320, Boca Raton FL 33431

and by CRC Press
4 Park Square, Milton Park, Abingdon, Oxon, OX14 4RN

CRC Press is an imprint of Taylor & Francis Group, LLC

© 2025 George Nadim Melhem, Paul Richard Munroe and Akshay Vithal

ISBN: 978-1-032-85169-3 (hbk)
ISBN: 978-1-032-85170-9 (pbk)
ISBN: 978-1-003-51690-3 (ebk)

DOI: 10.1201/9781003516903

Typeset in Times
by Newgen Publishing UK

Contents

Figures

Tables

About the Authors

George Nadim Melhem is currently an Adjunct Professor at the University of New South Wales (UNSW) in the School of Materials Science and Engineering, which is ranked 1st in Australia and 19th in the world (2024 QS World University Rankings). Holding a Bachelor of Metallurgical Engineering (Honours), Master of Materials Science and Engineering by research and PhD from UNSW, he additionally conducts research there. Professor Melhem has been a board member for the past decade contributing to the UNSW strategic direction for teaching and research. He lectured in Materials in Architecture, metallurgical courses, laboratory research and development as well as post-graduate electives for final year students at the university three decades ago. As the lead author and expert on aerospace materials (and on behalf of UNSW), Professor Melhem has published several papers in international journals and chapters in Encyclopaedias. He specialises in the use of aluminium alloys, most steel alloys, metal matrix composites/graphite epoxy composites in aerospace applications – including solving and designing for complex civil and structural applications in major infrastructures (through his own engineering company). Professor Melhem was formerly a principal contractor (designer and consultant) for QANTAS Airways on Boeing aircraft components. Currently, Professor Melhem holds post-nominals as a Fellow Professional Engineer of Australia (FIEAust) in the four disciplines (Aerospace, Civil, Mechanical, and Structural Engineering), is a Chartered Professional Engineer (CPEng), National Engineer Register (NER), Asia Pacific Economic Co-operation Engineer (APEC), is on the Register of International Professional Engineers (IntPE (Aus)) and is a Member of the Institute of Engineers (MIE). Globally, he is a panel interviewer for engineers and aerospace scientists; those seeking Chartered Professional Engineering Status (CPEng) with Engineers Australia. A Registered Professional Engineer of Queensland (RPEQ) on the Board of Professional Engineers of Queensland (BPEQ) also in the aforementioned four disciplines, he teaches and mentors students within his company (from five major universities in Australia) in these respective four disciplines. He has been a key contributor to commercial aircraft in aerospace materials science, structural engineering, tooling, and ground support equipment – and continues to service government sectors, providing design and consulting in major infrastructure projects involving civil and structural engineering.

Paul Richard Munroe is currently a Professor in the School of Materials Science and Engineering at the University of New South Wales (UNSW). Professor Munroe received his Bachelor of Science (Honours) and PhD in Metallurgy and Materials from the University of Birmingham, England. He is formerly the Deputy Dean for Research in the Faculty of Science at UNSW. Prior to that role, he was the Head of the School of Materials Science and Engineering – and, before that, Director and Professor of the Electron Microscope Unit. He taught (and continues to teach) a wide range of courses at UNSW, with his most significant contributions being in the field of microstructure-property relationships in advanced engineering materials. Areas of research conducted include functional thin films, intermetallic alloys, advanced metal matrix composites, thermal spray materials, surface modification of materials and biochars. He was one of the founders, as well as the Inaugural Technical Director, of the Australian Microscopy and Microanalysis Research Facility. A former Associate Editor for the journal "Materials Characterization", he has sat on a number of journal editorial boards and currently sits on the editorial board for the journal "Metals" – and served as a member of the Australian Research Council College of Experts. Professor Munroe has authored over 600 papers and provides advice to industry through his expertise in microstructure-property relationships.

Akshay Vithal is an Aerospace Engineer, currently working as the Aerial Design Lead for Geodrones Australia. He was a former student under the guidance and mentorship of Professor Melhem in his company in the field of Aerospace Engineering, researching several aspects of material and structural effects in aircraft in supersonic flight. Having graduated from the University of Sydney (USYD), Mr Vithal specialises in the aerodynamic design, flight performance, and structural analysis of various Uncrewed Aerial Systems – consisting of multi-rotors, fixed wing, and rotary-wing aircraft. Mr Vithal has experience with designing and building wing structures, wind tunnel testing, Computation Fluid Dynamics (CFD) modelling and testing, and Finite Element Analysis (FEA).

Preface

The aim of this book is to present to the reader an introduction to aerospace components such as rivets used in subsonic and supersonic aircraft. No detailed knowledge is required for the reader in either materials science or in structural dynamics to understand this book. Although, this book would be of benefit to university students in the field of materials science, structural engineering, and aerospace/aeronautical engineering. This book is also suitable for professional practicing materials scientists/metallurgical engineers, and aerospace/aeronautical engineers as it emphasises considerations of some of the effects of aerodynamic heating on aerospace rivets and various methods for testing these materials, which is not widely discussed in literature within the aerospace industry.

The book is divided into three chapters. Chapter 1 provides background into subsonic aircraft, and focuses primarily on subsonic aircraft with respect to the following: typical types of forces, case studies (earlier and past failures in aircraft, lessons learnt, and improvements made), design philosophies, selecting materials for commercial and military aircraft applications (optimum materials and materials based on weight reduction) – including the various alloys used to manufacture rivets/fasteners, and also the heat treatment of these alloys. Chapter 2 discusses the background on both commercial supersonic and military supersonic/hypersonic aircraft with respect to the following: selecting materials, rivet arrangement, skin friction/drag effects, estimating temperature, thermal properties, and fatigue testing of aerospace materials including rivets. Chapter 3 discusses the motivation for the selection of materials in the Concorde such as the rationale behind the structural design for the engine, aircraft skin, and the rivets. The design was based on the requirement of the aircraft to travel at a speed of Mach 2.0, and consideration for the varying aerodynamic factors affecting the engine, skin temperature, speculation of the rivet temperature, and possible structural effects due to heating and prolonged flight times.

Chapter 3 further discusses considerations for future research which would be undertaken in a laboratory such as optimal and practical means for both heating up and analysing the microstructure of the rivets, and extrapolating these microstructural findings to determine the possible effects of aerodynamic heating on aerospace rivets due to high Mach speeds. One method of analysis would be through the use of ANSYS Fluent software or through the use of SolidWorks Flow Visualization as they both admit Computational Fluid Dynamics (CFD) modelling. However, microstructural analysis is not possible with such programs, since the temperature is based upon the surface of the rivet only. Pronounced heating effects in localised regions of rivets in supersonic aircraft are commonly located in the wing sections, hence, supersonic wind tunnel testing would be the most practical means to achieve the simulated version of temperature variation across the rivet during supersonic travel and aerodynamic heating.

Microstructural changes in the rivet should be evident when subjected to wind tunnel testing. Preferably, the rivets selected would be close in chemical composition and mechanical properties to the rivets that were used in the Concorde. Although the precise type of rivets installed in the Concorde is not readily available in literature, and appears arcane, the microstructure of the rivets can be examined with a transmission electron microscope (TEM) prior to, and subsequent to, wind tunnel testing to observe the differences in the microstructure. In order to gain further understanding of the aerodynamic effects on rivets, the use of the Confocal Scanning Laser Microscopy (CSLM) is recommended, as it would highlight the temperature spread from the surface of a rivet to its centre which would represent in situ real-time phase transformations, and there is no requirement to cut the rivet. Further, a detailed discussion on how Vickers microhardness on the rivets could be converted to infer the yield and tensile strength of the rivets in lieu of conventional tensile testing, since most rivets would be too small to be tensile tested and the results would not be meaningful in any case.

The literature reviewed and presented in this book is by no means exhaustive or all-embracing, nor is any claim made towards relative importance with regards to literature cited over absent citations. Also, the data offered in this book should not be considered as permissible design, since the data presented are merely a general guide. The reader or designer must refer to approved design manuals which present data that have been derived statistically. A concerted attempt nonetheless has been made to include as much relevant information as possible based upon our past research, our publications, and the literature review to date in this field. It is our aim to highlight the pertinent factors in what we consider to be critical with respect to select materials that are used to assemble an aircraft, for instance typical connections such as rivets. Unlike traditional subsonic aircraft, the design of components for supersonic aircraft such as the Concorde have necessitated distinctive considerations of explicit material properties and manufacturing techniques in order to withstand the extreme conditions during cruise at supersonic speeds. Investigation of rivets under supersonic conditions could significantly increase the understanding of material microstructural processes which occur during these extreme conditions and, ultimately, the aim is to create safer flight conditions. Investigating supersonic conditions by utilising historic information and modern-day simulations would help to determine if the thermal variances in the flight cycle could affect the structural adequacy of the rivet and the structure it is connected to. The aim is to understand what is required of rivets and their connections, and how to better design for longevity in these materials after sustained use in flights approaching and speeding beyond Mach 2.0. This book presents some deliberations that would open the way for further considerations of both temperature and material characteristics, ultimately for a new generation of supersonic aircraft as the Concorde once afforded us.

Subsonic Aircraft

Forces and Suitable Materials for Rivets

1

1.1 VARIOUS FORCES ACTING ON SUBSONIC AIRCRAFT

Aircraft have been continually improving since their debut as a powered medium on the world stage in the early 20th century. Although, the Wright brothers were not the first to build or think about aircraft, they were the first to contribute seriously to aeronautical engineering. The Wright brothers originally experimented with glider type designs between the years 1900 and 1902, but they could not achieve enough lift. They continued and through their perseverance experimented with lift and drag, not only by flying, but also by setting up wind tunnels on hundreds of constructed wings precisely designed for their specific testing. They then progressed to powered design, and they considered many aspects of control such as roll and yaw. They worked on the wing and on a steerable rudder to compensate for the former control issues respectively. Surprisingly, they also did not, as others did, leave safety just to chance. They deliberately designed the low-powered engine to ensure low flying speed in order to fly into head wind for additional aid in lifting. Deservedly, the Wright brothers are accredited with being the first to attempt to solve the problem of control, power, and safety in aircraft.

Motivation for further development in earlier aircraft was immeasurable, but the spontaneous developments which pushed the envelope of design and construction of aircraft were due in large to factors correlating with the events of the time. For instance, the first aircraft between the time of the Wright brothers and World War I (WWI) were mainly designed and constructed to ensure strength. During WWI, strength, flying further, flying higher, and speed were of major importance. However, steadiness in the aircraft was also vital for a gun to be directed at its opponent in the sky. Later developments after WWI

DOI:10.1201/9781003516903-1

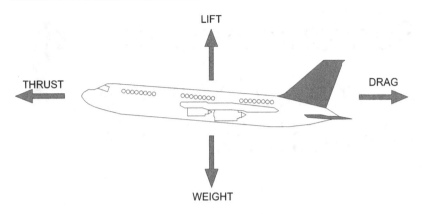

FIGURE 1.1   The Basic Forces on an Aircraft (Weight, Lift, Thrust, and Drag).

required sustainability and durability. Planes were no longer low-powered or made from wood, rather aluminium would replace wood and this was the emerging new material.

Each component in an aircraft is designed precisely to withstand a specific load or resist stress. Since the components of an aircraft have specifically been designed to carry out exact functions, it is useful to look at a few basic forces involved in aircraft (see **Figure 1.1**). There are generally five types of force imposed on aircraft, and these are compression, tension, bending, shear, and torsion. Compression is when material becomes squeezed and hence shortens the material. The upper section of the wings is placed in compression as the wing bends upwards to achieve uplift. Tension is the force that wants to pull a material apart and stretches it. An aircraft is placed in tension as the engine is thrusting it forward, whilst the air resistance drags the body backwards. Also, the lower part of the wing is placed in tension. Bending involves both compression and tension, that is, bending of the wings upwards, and shear is the resistance of material to slide upon one another. Often rivets in aircraft are subjected to shear. Torsion is a result of twisting, and this occurs in aircraft due to the engine imposing several types of forces on the airframe. Aircraft generally comprise of a fuselage, wings, and tailplane. The fuselage houses the passengers, air crew, cargo, weapons, and fuel. The wings are for lift, and the tailplane is for control of direction. The four primary control systems in an aircraft which also assist in keeping the aircraft steady are ailerons, rudders, elevators, and throttle. The ailerons located on the wings control the roll, the rudder located at the tail controls the yaw, the elevator located on the tail controls the pitch, and the throttle controls the thrust of the aircraft. Wing flaps located on the wings also allow the pilot to further control the lift of the aircraft as it takes off and lands.

1.2 INTRODUCTION TO SELECTING MATERIALS FOR AEROSPACE APPLICATIONS

The variables involved in the characterisation of fasteners for use in the aerospace are immense. The reason for this is that the assembly of aircraft parts is more involved than the assembly of most other structural components that we have on the ground. Since aircraft are subjected to a variety of loads in various directions, as discussed earlier, this necessitates many components of differing materials/strengths to be considered and used. With the advancement of the aircraft and its design, consideration must be given to design for a myriad of metals and composites combined. The challenge is to have the structure act in unity, which means that when all the loads transferred to each member (each material connected to another must be compatible so that corrosion is minimised or eliminated as much as possible), the size and configuration of the wing commonly known as the aspect ratio must be defined at the design stage so that the correct uplift, speed, and manoeuvrability is achieved, and so on.

The challenges faced in aircraft construction include, but are not limited to the following:

- incorrect design,
- lack of available information/knowledge/history,
- environmental factors,
- temperature,
- initial preload,
- fatigue,
- vibration,
- either incorrect maintenance scheduling or incomplete/incorrect maintenance,
- undetected corrosion hidden beneath complex joints that are not easily accessible or exposed to natural rain for washout of debris,
- dissimilar materials being joined to one another (metal/metal, metal/composite) causing galvanic corrosion,
- incorrect heat treatment of metals or incorrect construction of composites (fibre orientation or voids and defects within matrix),
- pilots incorrectly manoeuvring aircraft and subjecting new materials (composites) to stresses beyond their design capabilities,
- excessive speeds causing friction and heating of aircraft components,

- bird strike effects,
- lightning strike effects, and so on.

The list for manufacturing/construction/material combination/human error, is too exhaustive to mention in this book. However, in mentioning the above, it is clear to see that the topic of a specific material to be connected in an aircraft is critical to overcome the numerous challenges (for instance, determining a specific component's capability to perform satisfactorily in service – and, as intended). The designer would need to consider how the component will be affected under specific conditions such as environment, varying material in contact with another material, varying stress levels and temperature (to name a few), and, in turn, how these conditions pose serious deliberation during the design stage. There are many variables for the design engineer to contend with before finally deciding upon the best means for joining components together to form a structural component. One common method for assembling certain components together is by using fasteners. Bolts can be selected from a variety of grades such as steel (carbon steel), alloy steel, or stainless steel. Titanium and aluminium can also be used as fasteners in aerospace applications. However, aluminium is not as common in aerospace applications, as they do have very restricted use when compared to the prior mentioned grades. Not only is the strength of aluminium alloys low when compared to other fasteners that are manufactured from steel, titanium, or nickel (Ni) based steels, it also presents a major problem with respect to galvanic corrosion when in contact with material such as carbon fibre composites. The use of fasteners in aerospace for joining safety-critical structural components has been one of the major considerations in aircraft design. Steel has been used for fasteners in landing gears, wing root, and in several attachments of the wing, and this is due to steel's excellent strength, stiffness, and fatigue resistance properties. Unfortunately, the trade-off in steel's properties is that steel has high density and with certain coatings is limited in service due to temperature and other service condition effects. Solid rivets are mostly manufactured from aluminium, followed by titanium, A286 stainless steel, and plain carbon steels. The head is easily formed by bucking it with a pneumatic hammer which will not crack as they have cold-formed capabilities, and these solid rivets mostly have universal heads. Solid rivets should not be used for joining composites together as the rivets expand into an interference fit; it is highly likely that delamination will occur in the hole region [1]. The typical rivets in aerospace applications are shown in **Table 1.1**.

TABLE 1.1 Applications of Various Composite Materials with Various Fastener Types [2]

FASTENER TYPE	FASTENER GRADE	SURFACE COATING	COMPOSITE EPOXY-GRAPHITE	KEVLAR	FIBREGLASS	HONEYCOMB
Blind rivets[a]	5056 Aluminium	None	Not recommended[b]	Excellent[b]	Excellent[b]	c
	Monel	None	Good[b]	Excellent[b]	Excellent[b]	c
	A-286	Passivated	Good[b]	Excellent[b]	Excellent[b]	c
Blind bolts[d]	A-286	Passivated	Excellent[b]	Excellent[b]	Excellent	c
	Alloy steel	Cadmium	Not recommended[b]	Excellent[b]	Excellent	c
Pull-type lockbolts	Titanium	None	Excellent[e]	Excellent[f]	Excellent[f]	Good/not recommended[g]
Stump-type lockbolts	Titanium	None	Excellent[e]	Excellent[f]	Excellent[f]	Good/not recommended[g]
Asp fasteners	Alloy steel	Cadmium/nickel	Good[h]	Excellent	Excellent	Excellent
Pull-type lockbolts	7075 Aluminium	Anodized	Not recommended	Excellent	Excellent	Not recommended

Notes:
a Blind rivets with controlled shank expansion.
b Metallic structure on the backside.
c Performance in honeycomb should be substantiated by installation testing.
d Blind bolts are not shank expanding.
e Use flanged titanium collar.
f Fasteners can be used with flanged titanium collars or standard aluminium collars.
g Depending on the fastener design, check with manufacturer.
h Nickel-plated Asp only.

1.3 FASTENERS AND DIVERSE COMPONENTS THAT MAKE UP AN AIRCRAFT

Extrapolating details from existing smaller fasteners shows, for instance, that for a larger diameter fastener using identical material, manufacturing and joint systems will not be a correct technique. Larger diameter fasteners have a substantially decreased fatigue limit capability than smaller fasteners of the same material, manufacturing, and joint details. There seems to be some grey areas in literature in accurately defining fasteners, and partly so due to the lack of unification of many standards throughout the world. Although there does exist a great variety of fasteners, rivets are one such fastener type, comprised of a mandrel in sleeve (one-piece fastener) which is mechanically formed or upset from one side. For instance, blind fasteners from one side are easily accessed, while the other side that is not accessible is upset by explosive means [2]. Threaded fasteners are easily removed even after assembly without any damage to the fastener. Pin fasteners may be either tubular or solid and are used primarily where shear load is the dominant load. Composite structures require fasteners that may be used in combination with adhesives in order that the highly loaded joints are adequately improved. In some instances, if the adhesive has been processed to stringent quality control, it may be stronger or more effective than the rivets that are spread in certain locations, as it spreads the loads over a wider region of the assembly than the individual mechanical fasteners. Most of the fasteners mentioned in **Table 1.1** can be used in composite joints such as rivets, pins, bolts, and blind fasteners.

The popular adage, "A chain is only as strong as its weakest link", also applies in aircraft components that are assembled together. This adage can also be related to many aspects within engineering or designing an aircraft, for instance, where strength was originally paramount in the earlier days and very early stages of development of the aircraft. It was eventually realised that strength was increased at the expense of increasing weight and increasing corrosion, although corrosion was not really understood until decades later. It was later realised that other variables apart from strength were even more important for the durability of the aircraft. Therefore, the strength of materials like the example or adage used in the chain, could apply to fasteners or any other material in the aircraft. Whilst it is obvious that fasteners must be strong enough to combine certain materials in the aircraft together, particularly where there are either varying loads or temperatures that they must withstand, there are also other variables which affect the fasteners and other assembly parts. Examples include: corrosion, human error, manufacturing processing,

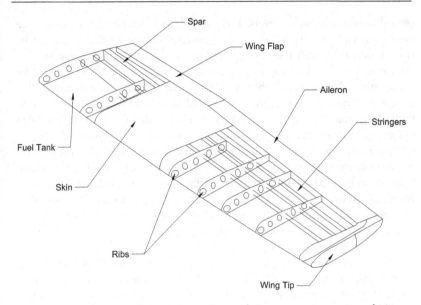

FIGURE 1.2   Schematic Showing Main Structural Components in an Aircraft Wing.

incomplete or incorrect design, design to eliminate stress raisers, metallurgical processing properties, stringent control on the construction/processing of new materials before reaching the assembly stage, galvanic effects due to incompatible material in connection with one another, and so on.

Fasteners are an integral component in the overall picture of assembling the structure, and millions of fasteners make up a great percentage of most aircraft. However, to understand how fasteners affect the overall structure, the members and components in an aircraft need to also be quantified, broken down, and defined so that their interaction with the fasteners are better understood. The structural components vary from fasteners to individual components such as stringers, ribs, chords, aluminium skins, composite components [3], and so on (see **Figure 1.2**).

1.4 CASE STUDY 1: EARLIER AIRCRAFT FAILURES DUE TO INCORRECT BASIC DESIGN

It is well-documented throughout engineering literature that materials generally fail at fluctuating loads at much reduced values of stress than that of their normal static failure stress levels. This particularly holds for aircraft also, since most – if not all – of an aircraft's components are subjected to fluctuating loads. A great deal of aircraft failures occurred due to fluctuating or fatigue loading even though attempts have been made to design and account for these. Fatigue is a continual degradation of the strength of a material during service, such that the material fails at a much lower stress level than the ultimate tensile strength level of that material. More than 80% of structures that fail by fatigue are due to the majority of materials undergoing cyclical loading; therefore, the damage tolerance principle is applied during the initial design stage. The damage tolerance criterion suggests that the material can withstand certain stresses whilst a defect up to a maximum size exists within the material [4]. The defects of course are to be regularly inspected via non-destructive means. The initiation of small cracks (micro cracks) due to fatigue can grow and propagate further under repeated or cyclical loading and eventually the material will fail if not detected and rectified early on.

Examples of three earlier fatigue failures on the English De Havilland Comet airliners took place on 2 May 1953, 10 January 1954, and 8 April 1954. Damage tolerance was recognised as a serious issue ever since these accidents first occurred in 1953. The latter two were confirmed to be due to the flight cabin pressurisation and depressurisation cycles that produce stresses such as longitudinal and circumferential stresses in the fuselage skin. The May 1953 failure was originally incorrectly thought to be due to a heavy tropical thunderstorm – but upon further assessment, confirmation of the second disaster in January 1954 and the third one in April 1954, attributed the same failure reasons to that of the failure that occurred in 1953. Interestingly, it was determined that even though the stresses involved were well below the allowable stresses for its cycle loading, the cause of the failures were due to stress concentrations emanating from the square corners of windows near the front cabin and around the rivets. Square corners must be avoided in design as they contribute to what is known as stress raisers, and rounded sections produce less stress and are therefore ideal wherever possible in design (see **Figure 1.3** and **Figure 1.4**). Due to repeated cycles of pressurisation and depressurisation, and ultimately the immense pressure, these cracks eventually progressed as fatigue cracks causing the fuselage to explode in the air at very high altitude.

After these unfortunate findings of metal fatigue which were first generated around the corners of the windows, all subsequent aircraft were redesigned with rounded windows and without rivets, thus mitigating as far as possible these unsolicited cracks. The "Fail-Safe" design (discussed further in section titled "Design Philosophies in Aircraft") approach was developed in the 1950s after the disasters, which meant that the assumption had to be that there is an initial defect size in every single material, even before it enters into service. Cracks can therefore increase to a certain size without compromising the safety of the aircraft, and monitored between flights. This is accompanied by

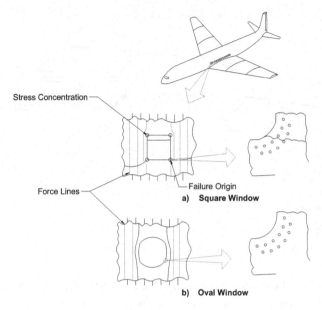

FIGURE 1.3 (a) Stress Concentrations Originating from the Square Corners of Windows Near the Front Cabin and Around the Rivets. (b) The Later Developed Oval Windows Have Much Reduced Stress Concentration in Comparison.

what is called crack propagation calculations at each given inspection, which predicts the size the crack will extend to, and the aim is to ensure that the crack critical size between each inspection is not reached. Also, window sections were later developed with a different type of reinforcement panel which was able to withstand a much greater resistance to fatigue failure.

1.5 CASE STUDY 2. VARIOUS FAILURE MECHANISMS OF AEROSPACE RIVETS

Rivets form a significant part of the structural components in the construction of aircraft, as they fasten aircraft skins to the internal substructure, hence they are used as a joining method [6]. As a result, rivets need to withstand mechanical loading and thermal stresses during the flight cycle (taxying, take-off, climb, cruise, descent, landing). However, due to these loads and stresses, the failure of rivets is a possibility. For example, mechanical cyclic loading can cause fatigue cracking which occurs around fastener (rivet) holes, as the

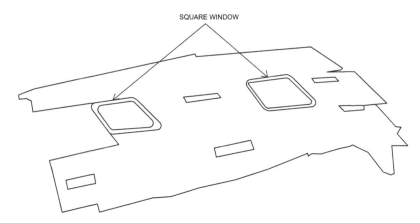

SQUARE WINDOW

FIGURE 1.4 Schematic of Roof Fragment of G-ALYP from the Fuselage of the Windows Which Failed on the 10th of January 1954 [5].

crack transfers through the material, and this can lead to the failure of the rivet. Furthermore, since rivets have an interference fit, aerodynamic heating – due to supersonic speeds – can cause the rivet to expand if it is of an incorrect grade, and as a result will fail. As the rivet expands, the mechanical, thermal, and physical properties change, which, in turn, drastically changes the capabilities of the material. When a material is stressed, it can exhibit accelerated corrosion, and this is known as stress corrosion. Stress corrosion of a material may cause the material to fail at approximately 10% of its typical ultimate stress level [7]. There are three principal modes of mechanical failure in which a riveted joint could fail, and these are: rivet shear, rivet/plate bearing failure, and net section failure [8]. However, failure due to corrosion in certain rivets of a particular grade may also be caused by the microstructure being altered via heat treatment or via heating effects as a result of supersonic speed. Of course, at such speeds, aerodynamic heating will define the type of material that will be used [7]. If the microstructure is altered by heating effects, then depending upon the material, the corrosion rate may also be affected. Single phase material may also have phases within their grain boundaries, which become more prone to corrosion, as certain grain boundaries may act as anodic sites, and the respective grains will become cathodic sites. Therefore, if the temperature of the metal is increased, some of the phases will both coarsen and potentially merge, which will also reduce the number of grain boundaries that act as galvanic cells, hence reducing the metal's overall resistance to corrosion. Whilst grain boundaries are generally anodic, in other instances, the grain boundaries may be cathodic sites. When only a single phase is present in aluminium alloys, for instance, typically these alloys will have a low corrosion rate [9]. However, when a second phase is present, the potential for corrosion is heightened. When these precipitates coalesce, the corrosion rate decreases, but not to the same extent as that of a single phase. The stress levels also impact upon the metal level of corrosion, and metals that have been cold worked are more highly stressed than those metals that have been annealed, and these stressed metals are also conducive to higher corrosion than that in annealed metals [8].

Design for dynamic and static loads in most structural systems is relatively straightforward compared to the design for impending rivet corrosion, as it is difficult to foresee most corrosion effects. For instance, there are factors that are inherent to the material in question, such as the rivet's microstructure and chemical composition must be considered, as they are both dependent upon the processing of the material. There are also factors that are not directly related to the material, but rather due to the environmental conditions which impact upon the rivets and upon the materials with which the rivets are associated with.

Fastening systems generally fail due to one or a combination of the following: 1) static loading, that is due to overload in tension, shear, bending or torsion, by 2) dynamic fatigue, which is cyclic loading (repeated loading) or by various forms of corrosion such as 3) galvanic corrosion, chemical corrosion, or stress corrosion [2]. The general areas of failure in most fasteners or rivets are directly beneath the head of the rivet, at the thread-shank in the bolt, or at regions of microstructural imperfections. Also, failure may occur due to the plates or sheets that are joined together experiencing dynamic fatigue [10]. Generally, fasteners are manufactured from a range of alloy grades and have protective coatings such as zinc, tin, cadmium or aluminium [2]. On the other hand, rivets are either manufactured from materials such as aluminium shank/ steel mandrel, all-aluminium or all-steel, and when combined with sheets of different materials, this creates a greater possibility for galvanic corrosion [11, 12]. Plates joined together, or a lap joint as shown in **Figure 1.5**, is the most basic construction methods used for joints, with the horizontal distance

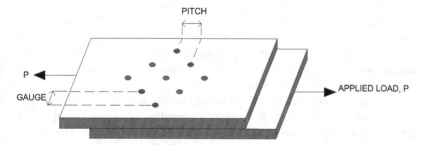

FIGURE 1.5   Lap Joint Showing Rivets or Bolts Penetrating Through Two Plates and Connecting These Plates Together.

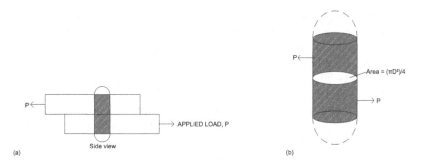

FIGURE 1.6 Side View of Lap Joint Showing Forces Applied Parallel to the Area in Shear.

between the holes, termed the pitch. **Figure 1.6** shows the rivet or bolt in the side view, and the bolt or rivet may fail across their diameters. **Figure 1.7** shows the plates in both elevation and plan view, and it can be seen from the top view on the schematic that when the loads are applied to the plates, the plate is actually pulled into the rivet. The plate immediately in contact with the rivet is placed in compression, and either the plate or rivet may fail in compression.

Galvanic corrosion is well-documented and understood [7, 13–16] and may cause failure of rivets. Different materials such as carbon fibre reinforced composites or graphite epoxy composites in contact with aluminium create galvanic corrosion. However, titanium will not cause galvanic corrosion with graphite epoxy composites [7]. Since metals are separated or ranked according to their electrode potential or metal behaviour which is defined under specific conditions such as in seawater, for instance, then the tendency for metals to reduce the ions of other metals below it in the series are defined by the elec-tromotive force (EMF) series (see **Figure 1.8**), thus making the difference in potential evident, ensuing in corrosion. The anode, cathode, and electrolyte must all be interconnected and act simultaneously for corrosion to occur, so that if only one of these factors is missing, then corrosion will not occur [17]. The surface area of the anode and cathode also plays an important role in corrosion of the electrodes, that is, the smaller the anode is in comparison to the cathode surface area, the greater the corrosion will be since the electrons move from the anode to the cathode.

Marine slimes occur in aircraft mainly in regions where diesel fuel and water on the aircraft have both resided, for instance, in regions on pipes and tanks as this encourages microbial growth to form the biofilm [17–19] in materials such as in the rivets and fasteners. Biofilm increases the rate of a chemical reaction which markedly increases the potential for corrosion.

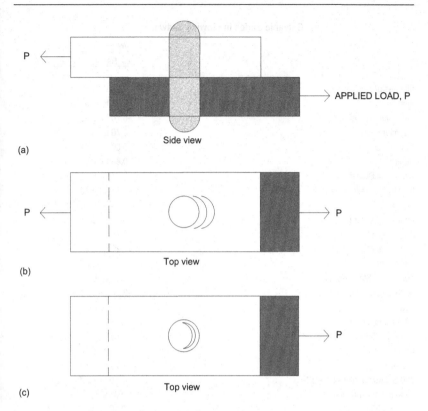

(a)

Side view

(b)

Top view

(c)

Top view

FIGURE 1.7 Illustration Showing (a) Section, (b) Compression into Sheet Adjacent Hole, and (c) Compression Being Induced Upon the Rivet Due to the Top Plate Being Pulled into the Rivet.

Aircraft have traditionally relied on rivets for the connection of the fuselage and wing skins, and they have stood the test of time when treated correctly with the appropriate design [20]. The lap joints are riveted and sealed by some manufacturers, whereas others employ a combination of riveting and adhesive bonding [21, 22]. Corrosion damage in the crevice geometry of lap joints is highly undesirable (see **Figure 1.9**). Fatigue cracking was present in the Aloha Boeing 737 aircraft, which occurred in 1988, in the large section of the upper fuselage (see **Figure 1.10**). This was due to a combination of factors. The cabin was subjected to repeated pressurisation cycles since the aircraft flew exclusively on short runs between the Hawaiian Islands, and the aircraft was subjected to high stress pressurisation during each cycle. As a result, this caused the formation of cracks at the fastener or rivet holes. Since the cracks were undetected by the maintenance crew, the cracks in the fuselage skin were enhanced due to the corrosion that existed in the joints. As a result, fatigue

Galvanic Series in Flowing Seawater

Alloy		Voltage Range of Alloy vs. Reference Electrode*
Magnesium	ANODIC or	-1.60 to -1.63
Zinc	ACTIVE END	-0.98 to -1.03
Aluminium Alloys		-0.70 to -0.90
Cadmium		-0.70 to -0.76
Cast Irons		-0.60 to -0.72
Steel		-0.60 to -0.70
Aluminum Bronze		-0.30 to -0.40
Red Brass, Yellow Brass, Naval Brass		-0.30 to -0.40
Copper		-0.28 to -0.36
Lead-Tin Solder (50/50)		-0.26 to -0.35
Admiralty Brass		-0.25 to -0.34
Manganese Bronze		-0.25 to -0.33
Silicon Bronze		-0.24 to -0.27
400 Series Stainless Steels**		-0.20 to -0.35
90-10 Copper-Nickel		-0.21 to -0.28
Lead		-0.19 to -0.25
70-30 Copper-Nickel		-0.13 to -0.22
17-4 PH Stainless Steel +		-0.10 to -0.20
Sliver		-0.09 to -0.14
Monel		-0.04 to -0.14
300 Series Stainless Steels ** +		+0.00 to -0.15
Titanium and Titanium Alloys +		+0.06 to -0.05
Inconel 625 +		+0.10 to -0.04
Hastelloy C-276 +		+0.10 to -0.04
Platinum +	CATHODIC or	+0.25 to +0.18
Graphite	NOBLE END	+0.30 to +0.20

* These numbers refer to a saturated calomel electrode.

** In low-velocity or poorly aerated water or inside crevices, these alloys may start to corrode and exhibit potentials near -0.50 V.

+ When covered with slime films of marine bacteria, these alloys may exhibit potentials from +0.30 to +0.40 V.

FIGURE 1.8 Galvanic Series in Flowing Seawater, Showing Voltage Differences Between Various Alloys.

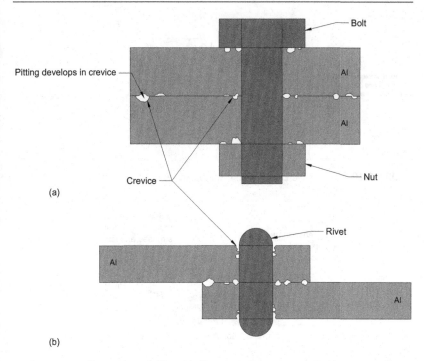

FIGURE 1.9 Illustration of Typical Crevice Corrosion: (a) Aluminium Sheets Connected by Bolt; (b) Sheet Connected by Rivet.

cracking was not anticipated to be a problem, provided that the overlapping fuselage panels remained firmly bonded together [11, 23].

However, a closer look at the corrosion processes in this crevice geometry reveals that, due to subsequent build-up of voluminous corrosion products inside the lap joints leads to pillowing, a critical condition is created whereby the overlapping surfaces are separated (see **Figure 1.11**). The predominant corrosion product identified in the corroded fuselage joints is aluminium oxide trihydrate, with a particularly high-volume expansion relative to aluminium. The build-up of voluminous corrosion products also leads to an undesirable increase in stress levels near critical fastener holes. Rivets have been known to fracture due to high tensile stresses resulting from pillowing [11, 24]. Corrosion damage on commercial and military aircraft, such as the pillowing in lap splices, is of major concern when referring to the ageing aircraft problem worldwide.

FIGURE 1.10   Schematic Description of the Aloha Aircraft Incident.

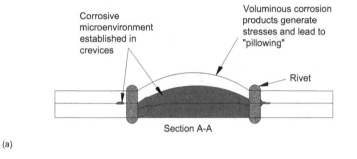

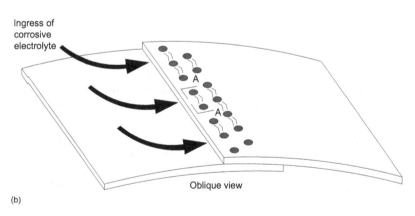

FIGURE 1.11   Pillowing of Lap Splices in Aircraft.

Tensile and fatigue strengths of rivets are lower than bolts of comparable diameter and machine screws with nuts. The allowable tensile loads of blind rivets of a certain diameter can exceed 1,045 kg. Rivets are fasteners that are the least susceptible to vibration loosening; for this reason, aircraft manufacturers prefer these types of rivets. Although riveted joints are weatherproof, they are not normally airtight or watertight under pressure. Therefore, a sealing compound, rivet coating, or special washer may optionally be used, but this is at an increased cost. The exactness of rivets that are manufactured in large quantities, cannot generally be expected to have the equal accuracy as that of screw-machine parts. Consideration should be given to rivets where the dimensional variation must be maintained as low as 0.025 mm [8].

1.6 CASE STUDY 3. AIRBUS A300-600 AND SPACE SHUTTLE COLUMBIA ACCIDENTS DUE TO COMPOSITE MATERIAL FAILURES

1.6.1 Airbus A300-600 Flight 587 Accident

On 12 November 2001, the vertical stabiliser (made from carbon/epoxy in an American Airbus A300-600 Flight 587) broke away from the fuselage during flight when it departed New York's John F. Kennedy International Airport (JFK): 260 individuals in the aircraft and five individuals on the ground died when the aircraft crashed in Queens, New York not long after take-off. The National Transportation Safety Board (NTSB) stated in their report that the captain and first officer were both certified, qualified, and that there were no fatigue or health issues. Traffic controllers who handled Flight 587 also complied with Federal Aviation Administration wake turbulence spacing (between aircraft Japan Airlines Flight 47 and Flight 587) requirements. With respect to the failure of the vertical stabiliser which is typically attached to the fuselage of an A300-600 aircraft, there are six attaching points or attachment lugs connecting the stabiliser and fuselage together. Each attaching point has two attachment lugs made from aluminium and two lugs made from composite material, and all of these lugs are connected via a titanium bolt (see **Figure 1.12** and **Figure 1.13** for the stabiliser, lug, and pin detail). The aluminium and titanium parts were not damaged in the crashed aircraft, but the composite material had failed. The description for this failure, as presented by the NTSB, was that the composite attachment lugs that were situated at the base of the fin failed. The reason for failure was a result of the first officer's overreaction; the officer applied large yaw control (pilot's overuse of the rudder) to correct the plane due to extreme turbulence. Excessive yaw control

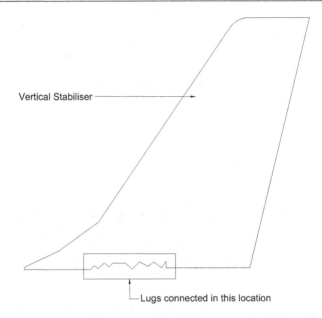

Vertical Stabiliser

Lugs connected in this location

FIGURE 1.12 Location of Lugs in Carbon/Epoxy Vertical Stabiliser of American Airbus A300-600 Flight 587.

input caused excessive bending in the attachment as there was exposure in the wake of a prior 747 aircraft taking off. Irrespective of the A300 tail fins being over-stressed above their design on several occasions in the past, the NTSB concluded in their report, that the composite material had failed since it was stressed past its design limit [25] which was due to the first officer's preventable and disproportionate amount of force applied on the rudder pedal. The failure began in the right rear lug, and the vertical stabiliser fully separated from the fuselage due to aerodynamic loading which exceeded twice the load design limit.

For this reason, the Australian Transport Safety Bureau (ATSB) report [27] states that a representative at Hawker de Havilland commented, as follows, "We have replaced the wing flaps on the C-130 Hercules Transport for the Royal Australian Air Force. These metal flaps normally start cracking after 3,000 hours. We stopped testing the composite flaps at 60,000 hours." The response of the ATSB to this comment in the report [27] was, as follows, "Nevertheless aircraft owners and operators should not be complacent about the durability of composite structures, and assume that they will never fail. Composite components can fail with catastrophic results, such as the crash of American Airlines Airbus A300-600 Flight 587 in November 2001...." [27].

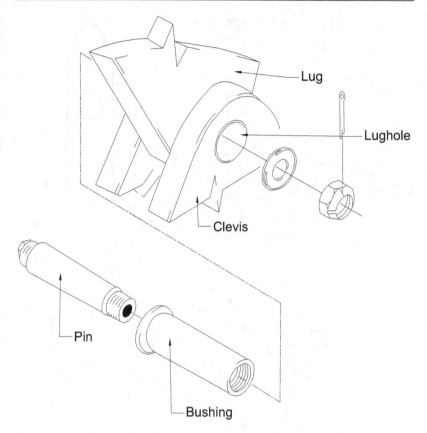

FIGURE 1.13 Lug and Pin Detail in Main Attachment Lug [26].

1.6.2 Space Shuttle Columbia Accident

On 1 February 2003, Space Shuttle Columbia after 17 days in orbit, began its re-entry into the earth's atmosphere. It was during this re-entry that the shuttle experienced excessive heating of the leading edge on the left wing that caused the catastrophe which led to the death of seven astronauts (see **Figure 1.14**). Temperatures typically exceed 1,650°C during re-entry, and this superheated air of oxidising plasma was sustained for up to 15 minutes. However, the breach actually occurred 17 days earlier (16 January 2003) at the Kennedy Space Centre, Florida, just after the shuttle took off. It was due to a large piece of insulating foam from the external tank which dislodged and hit the leading edge of the space shuttle's left wing on its underside. In a matter of two hours since losing the signal from the Columbia spacecraft, NASA's administrator

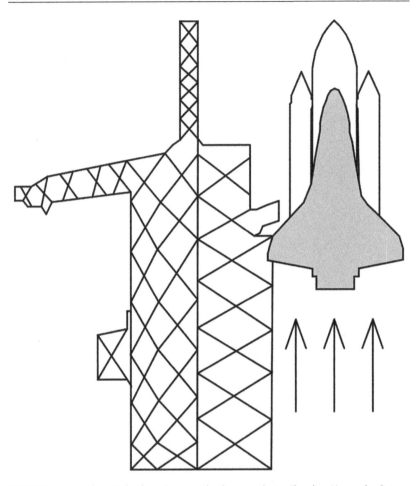

FIGURE 1.14 The Columbia Space Shuttle Launch in Florida, Kennedy Space Centre in Cape Canaveral, 2003 [28].

quickly formed the Columbia Accident Investigation Board (CAIB) and released their report seven months after the accident, on 26 August 2003. CAIB performed their own testing, and concluded that the falling foam from the external fuel tank (which contains cryogenic fuel for the orbiter's main engines) impacted the shuttle wing's leading edge and created a hole about 250 mm wide. Sensors were attached by NASA to the shuttle near the skin and within the insulation so that the astronauts within could obtain real-time data on the temperatures on the shuttle's surface, and so on. Thus, CAIB's investigation and testing detailed in their report that this impact from the foam would have created a hole in the composite material Reinforced Carbon-Carbon

(RCC) panel, only 82 seconds after take-off. This was not the first time that NASA learnt about foam dislodging and hitting the leading edge of the wing, as it has happened several times over many years. Despite constant warnings from the engineers in the past, the NASA management team downplayed this problem and treated it as non-critical to the mission. CAIB mentioned that NASA should have also learnt from the problems that they had faced with the Challenger disaster in 1986, and not repeat or ignore these problems. The RCC was designed to counter the highest temperatures that would be imposed upon the shuttle due to re-entry in regions such as the leading edges of the wings, the area aft of the nose cap, the forward orbiter, and the external tank. Upon re-entry, the extra heat from the superheated gases penetrated the wing where it was damaged, and began melting it from the inside, which led to the shuttle tearing apart and losing its crew.

Since temperatures imposed upon materials are very severe during re-entry, this highlights the unfortunate reality of design for materials, and that is every material has certain strength levels and thermal properties up to a certain point before final degradation. The CAIB report of 2003 critically pointed out that there were more factors than just the physical and technical causes that eventually led to this catastrophe. CAIB stated in this report that many other reports will identify the issue at hand (due to some component in the shuttle) as being broken or malfunctioning. In turn, they would place blame on the individual that was connected to that component, or lay blame on the engineer who has miscalculated in his/her design, or blame the operator who was responsible for missing signals or pressing the wrong switch, and so on [29]. Even though the CAIB did not understate the physical and technical causes, it emphatically spelled out the necessity that NASA must use more advanced, non-destructive testing on their materials in order to mitigate future risk to both crew and shuttles [30]. CAIB also turned their attention to highlighting the culture of NASA as being more concerned about being "successful" rather than being a "failure".

NASA has earned its success from the Apollo mission, and the phrase that has generally been adopted since that mission (though not formally), is "Failure is not an option", which basically represents the culture of "a can-do" approach within NASA. However, during a NASA press conference on 23 July 2003, a question was directed to a NASA personnel to answer, but no answer was received. The question was "If other people feared for their job if they bring things forward during a mission, how would you know that?" This is one example that highlights the fact that large organisations such as NASA, are more entrenched in a culture of success, and individuals feel there is no need to bring up matters that could allude to failure in any sense. The shuttle program in general was under immense pressure to meet austere deadlines, and there were budget constraints. Due to lack of overall funding to NASA for the shuttle program, the Bush administration ordered that the shuttle program end in 2010; however, the last shuttle flight was made by the space shuttle Atlantis on 21 July 2011.

CAIB reported that, had NASA realised the severity of the damage to the left wing by the foam earlier on, it would have had only one of two options to return the crew back home safely. The first option would have been to physically repair the damage on the left wing, or the second option would have been to send another shuttle into space to rescue the crew. NASA evaluated the first option as a "High-Risk", whilst the second option as "challenging but feasible". For a real measure of success, aviation must learn from instead of repeat mistakes and treat every single accident with the utmost solemnity. Management, engineers, assembly line workers, and everyone involved in aviation, must be encouraged to work collectively and be open to state any issues that could possibly be detrimental to the aircraft in their opinion – and, in turn, this must be encouraged and enacted by their superiors.

Although there have been some unfortunate accidents and major setbacks in aviation over the past several decades, it cannot be denied that there have been lessons learnt and mammoth improvements in general after each incident. The CAIB report, which formulated several recommendations for improving the safety of the shuttle including the management process, design criteria/ facets, and safety procedures, created a historical and long-term impression on the future travel in space. Aviation and space travel have lifted the bar to ever greater heights; this is the very reason general aircraft passengers today can safely concede that flying indeed is the safest form of travel.

Although the failures in the above case study for the A300 aircraft and the Columbia shuttle were primarily due to direct forces (prior history of impact from falling foam on shuttle wing *or* pilot overloading force on rudder to correct against turbulence) upon the composite materials in their respective uses, the reasons for these specific failures have been clearly outlined above, and these have been, in principle, due to human oversight or overreaction. Nonetheless, composites will continue to fail and have their limitations; they will not wane but will in fact be in more demand in the future, and be incorporated in greater quantities in both aeroplanes and space vehicles. The reason for this is composites are very light and have some superior properties over metals in many instances. Some of the main driving forces when considering aerospace travel are the cost of fuel continuously increasing, along with immense pressure to conform to improving various environmental factors such as reducing greenhouse gas emissions and lowering the overall carbon footprint. Composites do however have both advantages and disadvantages. Composites may be either natural or synthetic, where natural materials are renewable, but these also have either advantages or disadvantages.

The main advantages of composites are as follows:

Composite materials basically comprise of two or more materials, such that when they are combined, they provide a unifying material with properties exceeding those of the individual original materials. Composites can be tailor designed to have improved properties such as high strength to weight ratios, stiffness, and density. They also have the added advantage of having good chemical resistance, and for not being subject to corrosion in salt environments, they are able to maintain their dimensional stability, and have excellent fatigue resistance.

The main disadvantages with composites are as follows:

Composites are prone to erosion, impact damage, delamination and debonding, fibre fracture, moisture pick-up, environmental degradation, much lower conductivity than aluminium, lightning strike, and galvanic corrosion. When carbon or graphite composites are in contact with aluminium, one solution to counteract lightning strikes on composites is to incorporate on its surface a ply of conductive material or conductive paint, hence, protection against ultraviolet light will require specific paints and primers. Additionally, composites do not yield as metals do when subjected to regions of high stress and are not therefore ductile (rather they are brittle), but are anisotropic. Final properties are directly related to manufacturing processes, highly sensitive to notches, subject to random property variation, and very costly. Fibre composites upon shattering produce airborne needle-like fibres that may cause eye and skin irritation, embed in lungs when inhaled (some advanced fibres such as E-glass possibly have potential to produce pulmonary fibrosis and other diseases), and noxious gases will be generated from the composites upon fire during an accident.

However, if we are to temporarily reflect on the holistic use of composites in aircraft today, Boeing and Airbus are good examples to use, since both these aircraft manufacturers make widespread use of composites in the fuselage of their aircraft in order to reduce the use of fasteners and rivets. Elimination of fasteners and rivets contributes to considerable weight savings – but what is less commonly recognised is that the fuselage is still being joined by fasteners and rivets nonetheless. The reason for this is adhesive bonding does not provide the designers and engineers with confidence that bonding of the composite material will sustain the required loading to thwart failure should the respective component reach its critical design load.

1.7 DESIGN PHILOSOPHIES IN AIRCRAFT

Design philosophies present an imperative deliberation for a "Safe-Life" and "Fail-Safe" design in aircraft. A Factor of Safety (FS) can be represented as the ratio of critical failure value to actual calculated value [31] and is as follows: Load Carrying Capability of Structure (LCCS) / Expected Loading (EL). Therefore, if the LCCS is 4 and EL is 2, then its FS is 4 / 2 = 2. Typically, FS values between

1.3–1.5 are considered very low and this is due to the high level of confidence of the designer for that component. FS of 4 or even 5 are used in design, but this is due to a low level of confidence for the component by the designers. This possibly stems from one or a combination of the following: inadequate knowledge of manufacturing processes, material characteristics, and expectation of specific component for application. For composites, a FS of 3 in some instances [31] has been applied by designers due to their lack of confidence in its performance.

"Safe-Life" design is the first design philosophy used for aircraft and it was applied for the finite life of a component. The aim of the approach for "Safe-Life" is to assess an average or mean for the expected life of a component under service, and limit the component to only a fraction of this average expected life [32]. There is a large "scatter" in the results of fatigue testing, and this is the reason for the need to perform numerous component testing and obtain a mean life. The maximum allowable service life that ensures a low probability life for fatigue is achieved by dividing the mean life by a safety factor of often a value of 4 [33]. Therefore, this allows for a high factor of safety, and it is this philosophy which was responsible for many aircraft being inappropriately designed during the 1960s. Also, this design applies in instances where it is not feasible to have regular inspections on a component, and when it is not practical to inspect a defect in a component (as it is in service or flight). Therefore, the failure may take place before the aircraft lands due to the defect being undetected and has enlarged to the critical size during that single flight. For this reason, "Safe-Life" design is applied to components where it is critical to substitute low cost for more weight. The design intends that the prediction of risk to failure is small or zero. Therefore, components are retired before they fail, and in many instances, this can be costly since many components may have been retired before achieving their calculated or test life. However, it was wasteful to retire all aircraft fleet before the average life was up. The progression to the "Fail-Safe" design philosophy incorporates fatigue failures that may occur in service [32]. This design philosophy is no longer in use, except for instances such as in certain helicopter components and aircraft landing gear [33]. The concept of leak before break (LBB) is stringently used in the nuclear industry to ensure that any pipes that carry coolants for a power reactor will leak before creating a disastrous failure [34]. Cracks will develop in the pipes causing it to eventually fail, but cracks must be detected at an early stage before it grows to a particular size and create a major catastrophe. The "Safe-Life" design therefore does not have such a concept, but the damage tolerant and "Fail-Safe" design philosophies do have a similar design philosophy to LBB.

Damage tolerance resulted from the lessons learned after the famous Comet failures in the early 1950s. The limitations of the "Safe-Life" design have been compensated for in the damage tolerance design. This design goes with the supposition that there is an initial crack, and that the component can withstand further loads whilst the crack grows, but only for a period whereby the critical size is reached [33].

"Fail-Safe" design, as the name implies, is to fail in a safe manner. This design philosophy assumes that the component or structure, although consisting of some form of damage, may still be operated safely. The prerequisite for this is that the damage may grow or increase to a stipulated level before requiring a component or structure replacement. For instance, a structure may have a crack that will grow whilst in service, but this crack should be detected before it reaches a critical size, and this is achieved by performing regular inspections. However, a few of the simple solutions besides the non-destructive tests that are performed at regular intervals are load transfer, multiple load paths, and crack arrest system between components to avoid failures of the entire structure due to failure of single components [33, 35]. The point to this is that if a component failure occurs, the load is sustained by its closer members in order that total failure does not occur and it fails safely. Also, this failure of the single component needs to be found and repaired as the balance of the load will inevitably reduce the fatigue life of the remaining components [33].

1.8 SELECTING OPTIMUM MATERIALS FOR AEROSPACE APPLICATIONS

Whilst a myriad of factors must be evaluated when considering the design of an aircraft, safety, longevity, weight, and cost effectiveness are amongst the most important on the list. A typical aircraft airframe comprises approximately 70–80% of its unladen weight in aluminium. An aerospace engineer can choose from over 120,000 different types of materials for use in an airframe or engine, and this number is expected to continue to grow in the future. It is also estimated that there are less than 100 types of materials, that is, alloys, composites, ceramics, and polymers that are suitable for aerospace applications [36]. The main reason for this limited number is due purely to the stringent requirement for aircraft material to be lightweight, damage tolerant, durable, easily manufactured and cost effective. Most of the materials are therefore ruled out as a result of these mandatory properties.

Jules Verne, a French author and science fiction writer in the 19th century, was the first individual to mention in his book, *From the Earth to the Moon* (published in 1865), the possibility of taking humans to the moon in a hollow cannonball and launching it from Florida. Verne called the rocket *Columbiad*, and he said that there were three men inside the cannonball. Verne's story actually provided the catalyst for those who were ambitious at the time and future generations, particularly those who believed in pursuing the science and technology that would lead to such an outcome. One century later, rockets were launched from Florida (since Florida is closer to the equator and the rockets receive greater energy there). The reason for launching near the equator is that the rotation of the Earth is fastest at the equator, when compared to the North and South poles or between the poles and the equator. The distance from the equator to the centre of the Earth is longer than any other distance on the Earth's surface to its centre in the core (the Earth is not a perfect sphere, but rather an oblate spheroid). Therefore, the gravitational pull is less at the equator than any other position on the Earth, and rockets being launched in the vicinity of the equator will require less fuel to reach orbit. Apollo 11 launched their rocket in 1969 from Florida; they named the rocket *Columbiad* and placed three astronauts in the Apollo to commemorate Verne. The Wright brothers incorporated an aluminium engine crankcase in their wooden plane in 1903. Aluminium was also used in many aircraft during WWI, and the use of aluminium in aircraft increased during World War II (WWII), through to its incorporation in rockets in the 1960s. Aluminium alloy technology continues to evolve and be used efficiently and economically in commercial and military aircraft.

Once again, we return to how there are many areas of an aircraft where the components are specifically subjected to a range of loading/forces, temperature, and environment, and it is this knowledge that benefits the designer in choosing the most appropriate material for that specific application. For instance, the top of an aircraft wing is subjected to compression and the lower side is subjected to tension. **Figure 1.15** shows that 7075 aluminium alloy is used in the upper wing skins and stringers as they have a good combination of the following properties: Compressive Yield Strength (CYS), Modulus (E), Fatigue (FAT), Fracture Toughness (FT), Fatigue Crack Growth (FCG), and corrosion resistance. The earliest material used in aircraft in the early 20th century as mentioned above was from timber, as it has a high strength to weight ratio when used as a laminate structure. Timber is no longer used in aircraft since the main materials of choice in aircraft today range from aluminium, titanium, and composites. Whilst the trend for the use of aluminium has been gradually decreasing in civil transport airframes in the early 1990s through to the mid-1990s, it remained greater than 50% of the material used in the airframe. The slow but gradual replacement of aluminium for titanium and composites for certain components in aircraft seems inevitable due to the advantages of increased strength, increased stiffness, and a further reduction in weight. Two major shortcomings associated with these materials is the lack of toughness and the cost, when compared to aluminium.

Research in aerospace and accomplishments have been the main impetus responsible for further developing alloys in other sectors, primarily in buildings and automotives. Aluminium is not only abundant, but due to the broad research in the aerospace field, these materials have become astonishingly economical.

Fuselage Skin	: 2024–T3, 7075–T6, 7475–T6
Fuselage Stringers	: 7075–T6, 7075–T73, 7475–T76, 7150–T77
Fuselage Frames/Bulkheads	: 2024–T3, 7075–T6, 7050–T6
Wing Upper Skin	: 7075–T6, 7150–T6, 7055–T77
Wing Upper Stringers	: 7075–T6, 7150–T6, 7055–T77, 7150–T77
Wing Lower Skin	: 2024–T3, 7475–T73
Wing Lower Stringers	: 2024–T3, 7075–T6, 2224–T39
Wing Lower Panels	: 2024–T3, 7075–T6, 7175–T73
Ribs and Spars	: 2024–T3, 7010–T76, 7150–T77
Empennage (Tail)	: 2024–T3, 7075–T6, 7050–T76

Material Properties
Corrosion
CYS = Compressive Yield Strength
  E = Modulus
FAT = Fatigue
  ( ) = Important, but not critical, design requirement

FCG = Fatigue Crack Growth
  FT = Fracture Toughness
  SS = Shear Strength
  TS = Tensile Strength

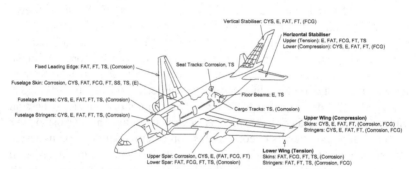

FIGURE 1.15 Material Selection for Structural Members of a Typical Passenger Aircraft [37].

Due to the beneficial properties of aluminium alloys such as light weight, high strength and corrosion resistance (to name a few), these materials make up most of the aircraft construction. Although aluminium alloys tensile strength and density (2.7 g/cm³) are low compared to steel, the strength to weight ratio and electrical and thermal conductivity are excellent.

Airframe materials comprise components such as the upper and lower wing, the fuselage landing gear, ribs frames, spars, and even the control surfaces. As mentioned above, the wings of an aircraft are subjected to tension on the underside and compression on the upper side of the wing. Both the upper and lower wing require high stiffness to density and yield strength to density ratios, and good resistance to stress corrosion cracking fracture or low fatigue crack growth rates. The fuselage is subjected to various forces (compression, tension, bending, torsion, and pressurisation forces) as it carries the entire payload. The fuselage also requires the same properties as the upper and lower wing in order to overcome the varying degrees of forces and potential corrosion cracking fractures.

Of particular importance also to aircraft is corrosion, and the control of corrosion plays a vital role in the structural integrity of the aircraft and ultimately the safety of humans. Since aircraft are exposed to a myriad of environmental factors including temperatures, it is imperative that consideration be given specifically to materials and coatings on aircraft to prolong the life of the components. Typically, commercial aircraft must remain in service for between 30,000–60,000 flight hours (2½–3 decades), whereas military aircraft range between 8,000–14,000 flight hours (5–12 years) [36].

When the aircraft is flying over the ocean or over the desert, the range in temperature may vary between -50 and 60ºC [38]. It is for this reason that composites are generally tested in extreme cases at temperatures of 71ºC with high humidity (85–95% water) to determine water absorption to simulate the worst possible condition for an aircraft [36]. Unfortunately, such tests can never simulate real-time exposure over the life of an aircraft, but generally materials can then be suitably classified in their respective roles for the expected service period. It was during the 1970s whereby the design for improved fatigue enabled the structure to be able to withstand certain critical damage (discussed in detail in "Design Philosophies in Aircraft" earlier). Following this, the 1980s experienced the development of design for prevention of corrosion where an understanding of material selection is central to mitigating corrosion. For instance, important factors to consider (as non-careful configuration in design may impart corrosion to these regions) are both the configuration and surface finish of the material in addition to the choosing of appropriate materials. Of significance is that the materials must be highly compatible with one another, in order to mitigate or eliminate galvanic corrosion or fluid entering between any two surfaces.

The material used mostly in an aircraft is aluminium alloys – and although aluminium as mentioned above has its advantages, it also has limitations. The main limitations of aluminium are strength and temperature capabilities when compared to other materials such as steel/metal alloys. Aluminium is not able to withstand temperatures above 660ºC, and it is this low melting point temperature which limits its use in maximum service temperature. It is reasonable then to segregate the application of the specific types of materials used in the engine and the main body or airframe of the aircraft since both have certain strength and temperature requirements. Turbine engines, for instance, typically use materials such as composite, steel, aluminium, titanium, and nickel. **Table 1.2** lists some of the common materials used in aircraft and their tensile properties. Of major importance in material selection in engines are the fluctuating operating temperatures. For instance, in the regions where the engine

TABLE 1.2 A List of Mechanical Properties for Different Materials Used for Aerospace Applications

MATERIAL	MASS DENSITY (g/cm³)	YIELD STRENGTH (MPa)	ULTIMATE TENSILE STRENGTH (MPa)	TYPICAL APPLICATION AND PROPERTIES IN AEROSPACE
Aluminium 7xxx (T6) forged (Zinc based)	2.7[36]	430–580[36]	530–620[36]	Highly stressed components; upper wing, stringers, and spars.
Aluminium 2xxx (T1) forged (Copper based)	2.7[36]	165–450[36]	300–480[36]	Structural and non-structural parts; fuselage skin, lower wing. Exhibits high fatigue resistance, strength and toughness.
Titanium (pure)	4.6[36]	172–483[40]	241–552[40]	Not heat treatable, but good weldability - rarely ever used in aircraft, but used for cryogenic purposes in fuel storage tanks.
Alpha-Titanium (α-Ti)	4.6[36]	760–960[36]	790–1030[36]	Not heat treatable, but good weldability - used in gas turbine engines due to its excellent creep resistance and ductility at high temperatures.
Beta-Titanium (β-Ti)	4.6[36]	1150–1300[36]	1000–1390[36]	Heat treatable and weldable, high tensile strength and fatigue resistance – used in SR-71 Blackbird Military, fuselage, wing, body skins, longerons, and ribs.
Alpha + Beta-Titanium (α + β-Ti)	4.6[36]	830–1300[36]	825–1580[36]	Heat treatable used in jet engines and airframes due to combined properties of α-Ti & β-Ti. High strength creep not as good as α-Ti.

(continued)

TABLE 1.2 (Continued)

MATERIAL	MASS DENSITY (g/cm^3)	YIELD STRENGTH (MPa)	ULTIMATE TENSILE STRENGTH (MPa)	TYPICAL APPLICATION AND PROPERTIES IN AEROSPACE
Titanium (Ti-6Al-4V)	4.5[36]	834–1013[40]	937–1186[40]	Heat treatable – used for high load requirements; aerospace fasteners, turbine engines, skins, wing box, stiffeners, and spars. Commonly referred to as "workhorse" of titanium is an α + β-Ti.
Magnesium	1.7[36]	110–200[36]	180–275[36]	No longer used much in aircraft due to poor corrosion resistance.
Carbon steel	7.8[36]	Low: 250–395[42] Medium: 305–900[42] High: 400–1155[42]	Low: 345–580[42] Medium: 410–1200[42] High: 550–1640[42]	Safety-critical structural components, such as steel fasteners for undercarriage (landing gear).
High strength low-alloy steel	7.8[36]	250–600[36]	Up to 2068[36]	Rarely used in aircraft – but are used in automobiles, trucks, and bridges.
Maraging steel	7.8[36]	660–2300[36]	970–2370[36]	Used in safety-critical structures in aerospace.
Stainless steel PH	7.5[43]	1150–1400[36]	1330–1517[36]	Engine pylons and certain structural components to withstand stress corrosion.
Carbon epoxy composite	1.6[36]	Brittle, do not deform before breaking[36]	3500–5500[36]	Jet engine and structural components.
Nickel superalloy	8.9[36]	900–1300[41]	1200–1600[41]	Jet engine part.

experiences high temperatures such as in the rear combustion sections, nickel-based superalloys and certain ceramics are used. However, titanium is used where there are low to medium temperatures, and in the lower temperature regions such as in the outer casing, either or both composite and aluminium are used. Generally, the specific strength of the material is lowered as a result of a higher working temperature. Turbine blades operate in the vicinity of 1,400°C, and high-speed aircraft or fighting aircraft flying at Mach 2.2 (Mach is defined as the ratio of aircraft speed divided by the speed of sound at sea level which is 340.3 ms^{-1}) and above, require the use of titanium alloys, particularly where the skin friction increases in temperature. The speed of sound in air is a function of temperature, and higher temperatures enable faster molecular collisions and faster propagation of the wave speed, however, the temperature of air decreases with an increase in altitude up to about 10 km.

Primary structures within aircraft can be affected by excessive external heat due to skin friction if not treated appropriately. Failures in aircraft are not necessarily related only to mechanical (fasteners, rivets, bolts, or adhesives) or chemical issues (corrosion, or incompatible material connections), but they are also largely related to degradation of components or structures via thermal processes. Aerodynamic heating plays a major role in the degradation of aircraft material, and the Lockheed SR-71 Blackbird Military's skin is manufactured from titanium for this reason. Military fighter aircraft use titanium alloys on a much larger scale, since in supersonic travel, the aerodynamic kinetic heating effects on the skin of the aircraft are more durable than any aluminium alloy, and this is why military aircraft use up to about 50% titanium alloys. Titanium was also used broadly as rivets and skin sheet in the aft end of the US F-15 fighter plane.

The effects of heating are first discussed in the section titled "Background on Supersonic Flight". Aircraft involved in hypersonic flight are designed such that their leading edges and panels maintain low temperatures in order to function without damage. For instance, use is made of convection cooling, by ensuring that the aerodynamic skin passes energy due to friction from the air in contact with the aircraft to a coolant that is of low temperature that travels through fins of a heat exchanger [39]. This in turn is directly connected to the primary structure of the aircraft, and for this reason convection cooling is imperative for protecting the primary structure from heating.

If ageing of materials is to be understood comprehensively, it would involve a long-term analysis ideally over the lifetime of the aircraft. However, results are required during the design stage, and it is not possible to have all of the details readily available. The main issue is that accelerated testing is not able to simulate real-time service exposure, and therefore it is difficult to predict the ageing of materials based upon time in any real sense, or for materials ageing via accelerated testing. As a result, prediction of the equivalent real-time behaviour with a high degree of confidence may be ruled out.

Since new materials continue to be further developed, it will be difficult even with the latest technology and latest experimental methods to predict the performance of material(s) based upon intended and expected lifetime. It is therefore imperative to have a system developed whereby an understanding of suitable testing and evaluation methods can better assist in the decision-making process. Selecting the most appropriate materials to withstand certain temperatures and structural requirements over a set period is key. The National Aeronautics and Space Administration (NASA) have requested the National Material Advisory Board (NMAB) assess these issues and have attempted to recommend accelerated evaluation methods which will provide some prediction for the robustness or longevity of aircraft materials and their structures over their lifetime. NMAB has also considered methods to assess geometry design which includes material manufacture, and assembly design such as fastening, welding, and bonding [44].

1.9 WEIGHT REDUCTION ALLOYS FOR AEROSPACE APPLICATIONS

Reduction in weight means reduction in overall weight and fuel consumption, and an increase in load carrying capabilities for aircraft. This was considered important during WWII since transportation by air, land, and water all depended upon this. The search was on for materials that would be low in density and have high strength to weight ratios. Less than a decade after WWII, these materials were extended towards space rockets and other advanced space technology. Nonferrous metals do not contain any iron and have other elements as their base and principal constituent. During the early 20th century (circa 1925), aluminium alloys were developed to a stage that they could replace timber in aircraft since greater strength can be achieved in the wings to counteract heavier loading and greater speeds for aircraft. We now turn to aluminium alloys.

1.9.1 Aluminium Alloys

The aluminium alloy that is mostly used in aircraft even to the present day is known as Duralumin. This alloy comprises of 93.5% aluminium, 4.4% copper, 1.5% manganese, and 0.6% magnesium. Professor Hugo Junkers from the University of Aachen in Germany was the first to implement the all-steel aircraft in 1914 which was found to be too heavy - and in 1915, he encouraged the use of Duralumin [45]. Loading on the wings from 1910 to 1920 was approximately 30–40 kg/m², but this was only suitable for timber frames, and subsequent generations of larger aircraft later had the loading on the wings increased further to between 500–1,000 kg/m², including increasing loads on the fuselage and other parts of the aircraft [36]. Timber lacked the stiffness that aircraft demanded, and steel was also lacking in that it was too heavy, so in the 1920s aluminium alloys were the material of choice for aircraft. The density of aluminium is 2.7 g/cm³, and has high specific stiffness and strength.

Aluminium in the commercially pure state is malleable, and is high in corrosion resistance since it only has a single phase. Aluminium in this state is not used where there is structural demand. When aluminium is alloyed with a certain percentage of other alloying elements, it can be tailor-made to suit certain strength and corrosion properties. Pure aluminium can be improved in strength via several methods such as adding alloying elements, cold working, or rolling. The tensile strength of aluminium can be increased by 500% by alloying with other specific elements (close to the tensile strength of steel), and the tensile strength of aluminium can be doubled simply by the physical processes of rolling or cold working. Whilst aluminium can attain strength comparable to steel, its weight is still maintained at approximately 30% that of steel. The other advantage of aluminium is that since it is malleable and ductile, it can be rolled into very thin sheets which are suitable for a myriad of aerospace applications. Aluminium alloys are based upon two main groups: wrought alloys and cast alloys. Wrought alloys are rolled, forged, or extruded – and hence can be shaped via these mechanical processes. Cast alloys are formed via investment casting or die casting. Cast alloys do not have any comparable mechanical properties to the wrought aluminium alloys, except that they can be formed in almost any shape desired.

The tensile strength and hardness of wrought aluminium is increased due to the addition of alloying elements via the process of solid solution hardening. Heat treatment on aluminium alloys will be discussed further below in the section titled "Effect of Heat Treatment on Aluminium Alloys". The main elements or metals that are added to aluminium include magnesium, manganese, copper, silicon, and zinc. The less frequently added elements are titanium, lithium, and nickel.

Magnesium is added to aluminium as it increases the strength and hardness through solid solution strengthening and increases its strain hardening ability. Manganese improves ductility and castability of aluminium when added to it. Copper significantly improves the tensile strength of aluminium by facilitating precipitation hardening and the formation of the compound $CuAl_2$, and increases the temperature properties and fluidity – but is poor in corrosion resistance and is susceptible to solidification cracking.

Silicon improves fluidity of the aluminium, and as a result it reduces the melting temperature of the aluminium alloy. Silicon is not alloyed alone with aluminium as it does not form an intermetallic compound with aluminium and cannot therefore have improved strength; rather, magnesium is added in order that it forms with silicon to form the intermetallic compound Mg_2Si which does improve the strength and hardness of the alloy via age hardening. Zinc achieves very high strength for the aluminium alloy, but is only useful in higher percentages as opposed to lower percentages and only when combined with other elements. Titanium is useful in refining the aluminium alloy and this also refines the weld microstructure within the filler wire which reduces weld cracking. The addition of lithium to aluminium lowers the density considerably, but the strength and Young's modulus is increased through precipitation hardening. Nickel enhances the temperature resistance, hardness, and strength when combined with aluminium-copper and aluminium-silicon alloys.

1.9.2 Aluminium-Lithium Alloys

Aluminium-lithium alloys belong to the 8000 series of aluminium alloys, and have been specifically developed for aerospace applications due to their attractive mechanical properties. The density of lithium is one fifth the density (0.53 g/cm³) of aluminium (2.7 g/cm³). As the weight percentage of lithium is increased in the aluminium alloys, the modulus and tensile strength increases linearly, but the density decreases linearly also. The level of increase in modulus is approximately double the decrease in density as the lithium is incrementally increased, thus yielding low weight for very large structures. When comparing these alloys to the aluminium alloys conventionally used in aerospace such as the 2000 and 7000 series, aluminium-lithium alloys have a superior resistance to fatigue. The disadvantages of aluminium-lithium alloys are low in toughness and ductility in the short transverse direction which often results in cracking. The processing cost is also high, but this is traded off when used for the purpose of light tanks in space shuttles where shuttle payload is increased as a result of large weight savings [36].

1.9.3 Titanium Alloys

Titanium was developed during the middle of the 20th century, and was primarily manufactured to counter the frictional effects and heat developed on aircraft skin due to supersonic speeds over Mach 2. Titanium alloys have an increased level of corrosion resistance when compared to most aluminium alloys, are rigid, have high strength, low density, and high fatigue resistance. These materials do not readily soften as most aluminium alloys would at speeds close to Mach 2, and instead can achieve at least 2 to 2½ times greater than this speed without any affect at all. Higher Mach speeds would necessitate the incorporation of more advanced titanium. The density of titanium (4.6 g/cm³) is higher than aluminium (2.7 g/cm³), and the cost is much higher than that for aluminium; also, the stiffness is not as high as in aluminium. Titanium can withstand similar loads to aluminium parts of greater size, but there is no weight reduction as a result of the smaller titanium part. The advantages these materials offer are more attractive in aerospace applications.

Pure titanium is not used in aerospace applications, except where there is a cryogenic requirement, such as in space vehicle fuel storage tanks that house liquid hydrogen. Titanium, under normal atmospheric pressure, can withstand temperatures down to −210°C and still retain its strength and toughness. Titanium exists in either pure form or one of the following alloys: alpha-titanium (α-Ti), beta-titanium (β-Ti), and alpha + beta-titanium (α + β-Ti). The titanium alloys have lower specific stiffness than most materials used in aerospace, and are limited to be used where there is no requirement for high stiffness in structural members. The alloying elements and heat treating of titanium can significantly improve the mechanical properties of titanium. Some of the titanium alloys that are weldable and heat treatable are summarised in **Table 1.2**. Ti-6Al-4V wrought was the standard alloy in which properties of all other titanium alloys were to be extrapolated from for comparative purposes, particularly when specifying a titanium alloy for a particular application in aerospace. It may be considered in any application where there is a requirement for high strength at moderate temperatures, and where excellent corrosion resistance and reduced weight are essential. Typical applications in aerospace for the commonly used titanium alloys are also summarised in **Table 1.2**.

The specific strength of titanium is better than most materials used in aircraft except for composites, and is therefore used where materials are subjected to high loads, that is, in the wing box, undercarriage/landing gear components, and airframe parts. Safety-critical parts such as the landing gear have conventionally been manufactured from steel due to the high strength, fatigue resistance, stiffness, and toughness. However, steels are susceptible to hydrogen embrittlement. Titanium is replacing steel as the landing gear, and is replacing many other types of metals. Other advantages include good fatigue resistance, good creep resistance at higher temperatures, resistance to exfoliation and stress corrosion cracking; it forms an oxide layer to repel most corrosive agents and is offered as a substitution to aluminium when excellent corrosion resistance is required, and has excellent oxidation for temperatures reaching up to 600°C [36].

In studying titanium and aluminium, to be manufactured into components for a particular application, the dimensions of the titanium components can be further reduced when compared with the same aluminium component. Titanium alloys may also be placed in an aircraft in regions requiring high temperature which may be considered too high for aluminium. Some common components for aerospace applications include jet engine discs, blades, casings, hatches, door edging, fastening brackets for wings, leading edge flaps, and also in several regions of the airframe. Since titanium is hard as a material, the general method to manufacture or repair it is by using a beam source. It focuses a beam that is high in density to reduce the metallurgical damage, since it has high processing speed and low overall heat input to minimise distortion. It has a rapid solidification rate in metallic components. Other aerospace metals that the laser is used on are aluminium, magnesium alloys, and high temperature nickel-based superalloys.

1.9.4 Magnesium Alloys

Magnesium for aircraft was developed during WWI by the Germans in 1915. Magnesium is one of the lightest structural metals available, with a density of 1.7 g/cm^3, and weighs about 25% that of steel and 66% that of aluminium. Magnesium has very low tensile strength, low ductility, and low resistance to both oxidation and corrosion. Interestingly, magnesium alloys have excellent strength to weight ratio, but are not suitable in structural applications. These very restrictions have been confronted by additional alloying elements and by age hardening. Aluminium, zinc, and manganese are the three primary alloying elements that are added to magnesium as they all have excellent solubility within magnesium. The different elements in varying amounts enhance certain properties to the magnesium alloy. For instance, zinc promotes an increase in strength in the magnesium, whilst simultaneously promoting a decrease in corrosion resistance. Aluminium in the same quantities as zinc may be added to counter the adverse effects, and also when aluminium and zinc are combined, the strength of the magnesium alloy increased. Manganese is a grain refiner and improves corrosion resistance.

Magnesium alloys, similar to aluminium alloys, are based either on wrought alloys and cast alloys. These alloys belong to either a heat treatable or non-heat treatable group. Magnesium alloys are either welded or riveted; however, the preferred method of joining is riveting, since galvanic corrosion via the welding process may become an issue due to magnesium being active and anodic. Aluminium alloy rivets are commonly used in riveting magnesium alloys since there is very little galvanic reaction (as they are very close to one another in the galvanic series ranking). Magnesium alloys have been used extensively on some wing sections, and parts of the fuselage, aileron, and wing tips. However, since the 1970s, the application of magnesium in aircraft has been declining ultimately due to its low corrosion resistance [36].

1.9.5 Effect of Heat Treatment on Aluminium Alloys

It is imperative to understand the effects of heat treatment in aluminium particularly when aluminium is to be welded where adhesive bonding, riveting, or bolting is not possible – and in instances where the alloy is to be used for structural purposes. The temperature also increases due to skin friction effects, as discussed earlier, influencing the microstructural and therefore the ultimate mechanical properties of the aluminium alloy. Heat treatment not only affects the structural properties via a decrease in strength, but it also has an effect on the corrosion resistance. However, if a material of medium to low strength is required, then non-heat treatable alloys would be suitable as they are also not affected by subsequent heat treatments and are much easier to fabricate. There are mainly two types of classes for heat treatment for wrought aluminium and wrought aluminium alloys, and these are the non-heat treatable alloys and the heat treatable alloys. Alloys that are non-heat treatable are those that require cold working to improve the mechanical properties. Heat treatable aluminium alloys rely on heat treatment to improve their mechanical properties. This involves heating to a required temperature in order that when it is held at this temperature for long enough, the alloying elements can enter a supersaturated solid solution (SSSS). The temperature is reached whereby only the primary phase exists, and the temperature is held long enough until the secondary phase has dissolved completely within the primary phase. The alloy is then rapidly quenched to suppress the secondary phase from leaving the primary phase in order to achieve an increase in strength. This condition is not stable at all, so the metal is then naturally aged at room temperature, otherwise known as age hardening or precipitation hardening. Also, an alternative ageing known as artificial ageing is achieved by heating at an increased temperature (typically 200°C).

Non-heat treatable wrought aluminium alloys do not respond well to heat treatment and therefore do not strengthen in this manner, and these alloys are either magnesium or manganese-based alloying elements. The strength is mainly improved via strain hardening which involves cold working during fabrication combined with dispersion hardening. The strength of non-heat treatable alloys cannot achieve strengths as high as the heat treatable alloys. The following lists the wrought heat treatable and non-heat treatable alloys along with their principal alloying elements and the designations are based on a four-digit index system: 1xxx, 2xxx, 3xxx, 4xxx, 5xxx, 6xxx, and 7xxx. The 1xxx,

3xxx, and 5xxx are non-heat treatable. The 2xxx, 6xxx, and 7xxx are heat treatable, and the 4xxx series consists of both heat treatable and non-heat treatable alloys.

In these designations, the first digit identifies the type of alloy it is or the principal alloying element, whilst the second digit identifies the alloy modification. If the second digit is zero, then the alloy is unchanged. The third and fourth digits are arbitrary numbers that designate the sequence of the alloy within the series. The 1xxx or 1000 series represents 99 wt% aluminium or higher, and has excellent corrosion resistance, increased electrical conductivity, increased thermal conductivity and workability, are readily weldable – but have reduced mechanical properties. The 2000 series has copper as its principal alloying element and has poor corrosion resistance when it is not clad with another aluminium alloy of higher purity, but its strength is comparable to that of mild steel. The 2xxx series is successfully welded if used in combination with high strength 2xxx series filler alloys. The 3000 series have manganese as the principal alloying element and is of medium strength, but have high temperature resistance capabilities. The 4000 series has silicon as the principal alloying element and is typically used in brazing or fusion welding and mostly this series is used as a filler material. The 5000 series has magnesium as the principal alloying element, has good corrosion resistance, is weldable, and is used in aircraft, buildings, bridges, and ships. The 6000 series has silicon and magnesium as its principal alloying elements, is easily worked and formed, has moderate strength and good corrosion resistance, is weldable and used in aircraft and building applications for structural components. The magnesium and silicon combine to produce the intermetallic compound magnesium-silicide (Mg_2Si). Zinc is the principal alloying element in the 7000 series, and the addition of either magnesium or copper facilitates precipitation hardening, making it one of the highest strength aluminium alloys [37, 46–48]. The 7xxx series alloys are often used in high performance applications such as aerospace, armoured vehicles, and sporting equipment.

There has been a universal attempt, particularly by ISO, to unify the nomenclature for all aluminium alloys. However, this has not been successful to date since each country uses its own nomenclature. This book uses the designations from the US Aluminum Association (AA), which uses the letter designations for different tempers applied to both wrought and cast aluminium [49–51]. Generally, the temper designation is placed after the alloy designation with a dash – and these indicate the sequence of either mechanical, thermal treatments – or both, used for various tempers (F is As-fabricated alloy with no further heat treatment or working of any sort, O is for wrought alloys that have been annealed to lower strength, H is for wrought alloys that have been strain hardened to increase strength, W is for naturally aged alloys and T is for stable alloys that are thermally treated with or without strain hardening).

For instance, alloys with a T followed by a number from 1–10 represent a specific sequence in the treatment procedure. T6 is a solution-treated and artificially aged alloy, and the mechanical properties of this alloy can be improved by precipitation heat treatment. This ageing treatment and sequence is well established based upon transmission electron microscopy (TEM) observations and electrical resistivity studies [47]. The sequence of ageing in 6061-T6, for instance, implies that following quenching the alloying elements remain in solid solution. Since these dissolved elements, as discussed earlier, are in a metastable condition, the alloy will decompose at a slow rate at room temperature. Hardening is maximised in these alloys when there is a critical dispersion of Guinier-Preston (GP) zones, intermediate precipitates, or both. Occasionally, the alloys can be cold worked after quenching and just prior to ageing. This method allows an increase in the dislocation density and it yields more sites at which heterogeneous nucleation of intermediate precipitates may occur. Since the GP zones are coherent or partially coherent with the matrix, it is these strain fields which produce the hardening. Overageing produces noncoherent zones and hence less hardening.

Aluminium alloys provided by the manufacturer have specific properties, and therefore when subjected to further treatment, may alter the original properties of the alloy. For instance, further heat treatment or welding on a specific region on the alloy will alter the microstructural properties of the material, and subsequently may affect and reduce either the strength or corrosion resistance properties. Welding causes a fusion zone which affects the microstructure in the vicinity, and differs from that of the microstructure of the parent alloy. For instance, 6061 aluminium alloy, which is an Al-Mg-Si alloy is the easiest to weld [37] and is not affected. Alloy 6063 is affected negligibly, but the copper alloys can be severely affected [52]. For this reason, Al-Mg-Si alloys (6xxx series) are used largely for structural and welded components in aerospace applications [53]. Filler metal for welding should ideally be as close as possible in solution potential to that of the solution potential in the parent alloy being welded. As mentioned earlier, with regards to galvanic corrosion, it is also important to ensure that the filler material being deposited onto the parent metal be cathodic (smaller region), whilst the parent alloy must be anodic (larger region) in order to mitigate corrosion. If the converse were true, then rapid corrosion would occur.

1.9.6 Alloy Rivets Used in Aerospace Applications

Subsonic aerospace rivets traditionally are manufactured from 7xxx series aluminium alloys, which contain Mg and Zn as the primary alloying materials

[11, 54]. The T6 temper applied to alloys, such as AA7075-T6, involves solution heat treatment, quenching, and ageing (which leads to an increase of the tensile strength of the alloy to >500 MPa).

However, at Mach 2.04, the temperatures experienced by a Concorde nose were estimated to be approximately 120–130°C which would have made AA7075-T6 susceptible to overageing [8, 55], thereby significantly reducing the overall tensile strength by up to approximately 40% [56]. The T6 temper increases the strength through the process of formation of a supersaturated solid solution during heat treatment, the formation of coherent Guinier-Preston zones during early ageing, and the nucleation of semi-coherent S' precipitates (Al_2Cu_1Mg) during late ageing [8]. When overageing occurs, the degree of coherency deteriorates and precipitation-free zones form, which decrease the resistance to slip and hence decrease the strength.

Owing to the temperatures experienced during supersonic flight, conventional rivets based on Al alloys could not be employed in the Concorde. Consequently, more temperature-resistant Ti alloys and Ni-based (viz., Monel) superalloys appear to have been used as rivet materials in the Concorde, although specific details are scarce [2]. Concerning such rivet alloys, there are three principal considerations: (1) microstructural alteration, (2) weight, and (3) oxidation. Further, beyond the considerations about the rivet materials, there are other considerations about the interactions of these materials with the Hiduminium RR58 aluminium alloy used for the skin of the Concorde. Rolls Royce developed this alloy as it will not overage during its life, and is one of the main structural components of the aircraft. The alloy is based upon an aluminium base composition with copper, magnesium, nickel, and iron to allow for elevated temperatures, fatigue, and creep. The three principal issues are the potential for (1) chemical reaction, (2) mutual solid solubility, and (3) thermal expansion mismatch.

From room temperature and up to 100°C, titanium has a lower coefficient of thermal expansion (CTE) of 8.40–8.60 μm/m°C [57] when compared to the Monel alloys which have a CTE of 14.0 μm/m°C [58], and titanium's density (4,540 kg/m^3) [59] is about half the density of Monel (8,840 kg/m^3) [58]. It would therefore be appropriate to utilise Monel alloys sparingly in only heat-sensitive areas, particularly in the vicinity of the engines. The downside to using titanium alloys is that they have a significantly higher specific gravity, which makes these titanium alloys heavier, and they are also more expensive to fabricate than their aluminium counterparts. These alloys, nonetheless, impart extra weight and extra cost to the aircraft. This choice deviated significantly from the original design criteria, since it stipulated a high strength to density ratio. The density of titanium is approximately 60% the density of steel (the density of steel is 7,850 kg/m^3). This compromise for the increase in density of titanium, as opposed to using lower density material such as aluminium rivets, is a measured trade-off for practical reasons, since titanium is better suited to tolerate a higher temperature environment and aluminium alloy rivets are

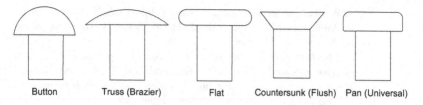

| Button | Truss (Brazier) | Flat | Countersunk (Flush) | Pan (Universal) |

FIGURE 1.16 Typical Rivet Heads [11].

not as galvanically compatible with carbon composite structures as titanium alloys [11]. Further information on the CTE of aluminium and titanium alloys is provided in **Table 2.1**.

A critical design factor in an aircraft's payload is the total weight of rivets used in the aircraft. The maximal take-off weight (MTOW) of the Concorde was 185,070 kg [60]. Titanium alloys are more difficult to machine than aluminium alloys, and therefore are much more costly to produce. However, titanium ranks highly when considered for certain applications on an aircraft when compared to most materials, since, overall, it has high corrosion resistance, relatively low density, and high strength [11] as the titanium rivets would be smaller in size than the aluminium rivets being used for the same application.

Fasteners that are commonly used for the assembly of aerospace components are categorised into various groups. For instance, blind fasteners are different from rivets since only one side is easily accessible [2]. Threaded fasteners may be easily detached even during subsequent assemblage without any damage to the fastener. Where shear is the dominant load type in the structure, pin fasteners with shapes such as tubes or rods would be preferable. Titanium fasteners exhibit high strengths, generally in the range of 965–1,380 MPa in tensile strength, and between 620–830 MPa in shear strength [61]. Much of an aircraft's individual structures demand fasteners to be combined with adhesives. Also, most of the fasteners can be used in composite joints such as pins, bolts, and blind fasteners as shown in **Table 1.1**. Rivets, however, are not recommended for joining composites together, as they will expand into an interference fit, which will cause damage around the hole by delamination. Typical head designs for rivets are shown in **Figure 1.16** below, and most rivets generally have pan (universal) heads.

REFERENCES

1   Barrett RT. *Fastener Design Manual*. NASA Reference Publication 1228. NASA National Aeronautics and Space Administration, Scientific and Technical Information Office, Lewis Research Center Cleveland, Ohio, 1990, p 29.

2   Jensen WJ. Failures of Mechanical Fasteners. In: *Metals Handbook, Vol. 11 Failure Analysis and Prevention*. 9th ed. Metals Park, OH: American Society for Metals, 1986, pp. 529–549.

3   Jernell LS. *Preliminary Study of a Large Span-Distributed-Load Flying-Wing Cargo Airplane Concept*. NASA Technical Paper 1158. NASA National Aeronautics and Space Administration, Scientific and Technical Information Office, Langley Research Center Hampton, Virginia, 1978.

4   Moreira PMGP, Silva LFM, and Castro PMST. *Structural Connections for Lightweight Metallic Structures*. Advanced Structured Materials. Berlin, Heidelberg: Springer-Verlag, 2012, p. 247.

5   Farag MM. *Materials and Process Selection for Engineering Design*, 4th ed. Boca Raton, FL: CRC Press, 2021, p. 49.

6   Wright B, Dyer WE and Martin R. *Flight Construction and Maintenance*. Chicago: American Technical Society, 1941, pp. 137–138.

7   Raymer DP. *Aircraft Design: A Conceptual Approach*. 6th ed. Reston, VA: AIAA, 2018, pp. 514–520.

8   Melhem GN. Design Variables for Steel and Aluminium in High-Rise Rooftops. PhD Thesis, University of New South Wales, Australia, 2008.

9   Vlack LHV. *Elements of Materials Science and Engineering*. 6th ed. Upper Saddle River, New Jersey: Prentice Hall, 1989, p. 534.

10 Forrest PG. *Fatigue of Metals*. Oxford: Pergamon Press, 1962.

11 Melhem GN. Aerospace Fasteners: Use in Structural Applications. Encyclopedia of Aluminium and Its Alloys. In: Totten GE, Tiryakioglu M and Kessler O (eds.), *Encyclopedia of Aluminium and Its Alloys*, Vol. 1. Boca Raton, FL: CRC Press, 2019, pp. 30–45.

12 Evans UR. An Outline of Corrosion Mechanisms, Including the Electrochemical Theory. In: Uhlig HH (ed.), *The Corrosion Book*. London: Wiley, 1948, pp. 3–11.

13 Marek MI, Natalie CA and Piron DL. Thermodynamics of Aqueous Corrosion. In: Davis JR (ed.), *Metals Handbook*, Vol. 13: Corrosion. 9th ed. Materials Park, OH: American Society for Metals, 1987, pp. 18–28.

14 Shoesmith DW. Kinetics of Aqueous Corrosion. In: Davis JR (ed.), *Metals Handbook*, Vol. 13: Corrosion. 9th ed. Materials Park, OH: American Society for Metals 1987, pp. 29–36.

15 Silverman DC and Puyear RB. Effects of Environmental Variables on Aqueous Corrosion. In: Davis JR (ed.), *Metals Handbook*, Vol. 13: Corrosion. 9th ed. Materials Park, OH: American Society for Metals 1987, pp. 37–44.

16 Shoesmith DW. Effects of Metallurgical Variables on Aqueous Corrosion. In: Davis JR (ed.), *Metals Handbook*, Vol. 13: Corrosion. 9th ed. Materials Park, OH: American Society for Metals 1987, pp. 45–49.

17 Melhem GN, Bandyopadhyay S and Sorrell CC. Use of Aerospace Fasteners in Mechanical and Structural Applications. *Annals of Materials Science & Engineering*, 2014, 1(4), pp. 1–5.

18 Videla HA. *Manual of Biocorrosion*. 1st ed. Boca Raton, FL: CRC Press, 1996.

19 Davis JR. *Corrosion of Aluminium and Aluminium Alloys*. Materials Park, OH: ASM International, 1999, pp. 151–153.

20   Ganguly S, Stelmukh V, Fitzpatrick ME, et al. Use of Neutron and Synchrotron X-ray Diffraction for Non-Destructive Evaluation of Weld Residual Stresses in Aluminium Alloys. *Journal of Neutron Research*, 12(1–3), 2004, pp. 225–231.

21   Roberge PR. *Handbook of Corrosion Engineering*. 2nd ed. New York: McGraw Hill, 2000, p. 4.

22   Komorowski JP, Krishnakumar S, Gould RW, et al. Double Pass Retroreflection for Corrosion Detection in Aircraft Structures. *Materials Evaluation*, 54, 1996, p. 80.

23   Wildey JF II. Aging Aircraft. *Materials Performance*, 1990, p. 80.

24   Komorowski JP, Bellinger NC, Gould RW, et al. Quantification of Corrosion in Aircraft Structures with Double Pass Retroreflection. *Canadian Aeronautics and Space Journal*, 42(2), 1996, p. 76.

25   Wang ASD. Fracture Mechanics of Sublaminate Cracks in Composite Materials. *Journal of Composites, Technology and Research*, 6(2), 1984, pp. 76–82.

26   *In-Flight Separation of Vertical Stabilizer, American Airlines Flight 587, Airbus Industrie A300-605R, N14053, Belle Harbor, New York, November 12, 2001*. Aircraft Accident Report NTSB/AAR-04/04. Washington, DC: National Transportation Safety Board, 2004.

27   *Fibre Composite Aircraft – Capability and Safety*. ATSB Transport Safety Investigation Report. Aviation Research and Analysis Report – AR-2007-021 Final. Australian Capital Territory: Australian Transport Safety Bureau, 2008, p. 21.

28   *Photography Index*. Washington, DC: NASA, Public Affairs Division, 1987.

29   Dick SJ and Launius RD. *Critical Issues in the History of Spaceflight*. Washington, DC: National Aeronautics and Space Administration, 2006.

30   The CAIB Report Volume 1 August 2003, Columbia Accident Investigation Board, National Aeronautics and Space Administration and the Government Printing Office, Washington, D.C. 2003, p. 25.

31   Shigley JE and Mischke CR. *Mechanical Engineering Design*. 6th ed. New York: McGraw-Hill, 2001, pp. 22–30.

32   Edwards PR. Full-Scale Fatigue Testing of Aircraft Structures. In: Marsh KJ (ed.), *Full-Scale Fatigue Testing of Components and Structures*. East Kilbride, Butterworths Glasgow, UK: National Engineering Laboratory, 1988, p. 23.

33   Grandt Jr AF. *Fundamentals of Structural Integrity: Damage Tolerant Design and Nondestructive Evaluation*. Hoboken, NJ: Wiley, 2004, pp. 19–20.

34   International Atomic Energy Agency. *Applicability of the Leak before Break Concept Report of the IAEA Extra Budgetary Programme on the Safety of WWER-440 Model 230 Nuclear Power Plants Status Report on a Generic Safety Issue*. Vienna, Austria: IAEA, 1993, p. 7.

35   Stephens RI, Fatemi A, Stephens RR, et al. *Metal Fatigue in Engineering*. 2nd ed. New York City, NY: Wiley-Interscience, 2001.

36   Mouritz A. *Introduction to Aerospace Materials*. Oxford: Woodhead Publishing, 2012.

37   Polmear IJ. *Metallurgy of Light Metals*. London: Edward Arnold, 1995.

38   Russell JA. *Industrial Operations under Extremes of Weather.* Meterological Monographs, Vol. 2. Boston, MA: American Meteorological Society, 1957.

39   Thornton EA and Wieting AR. Finite Element Methodology for Transient Conduction/Forced Convection Thermal Analysis. In: Olstad WB (ed.), *Heat Transfer, Thermal Control and Heat Pipes.* Reston, VA: American Institute of Aeronautics and Astronautics, 1980, pp. 77–103.

40   Donachie MJ. *Titanium. A Technical Guide.* 2nd ed. Materials Park, OH: ASM International, 2000, pp. 97, 101.

41   Satyanarayana DVV and Eswara Prasad N. Aerospace Materials and Material Technologies. *Aerospace Materials*, 1, 2017, p. 215.

42   *Materials Data Book.* Cambridge, England: Cambridge University Engineering Department, 2003, p. 12.

43   Davis JR. *Stainless Steels.* Materials Park, OH: ASM International, 1994, p. 104.

44   Committee on Evaluation of Long-Term Aging of Materials and Structures Using Accelerated Test Methods. *Accelerated Aging of Materials and Structures, the Effects of Long-Term Elevated-Temperature Exposure.* Washington, DC: National Academy Press, 1996.

45   Teichmann FK. *Fundamentals of Aircraft Structural Analysis.* New York: Hayden Book, 1968.

46   Avner SH. *Introduction to Physical Metallurgy.* New York: McGraw-Hill, 1974.

47   Hatch JE. *Aluminium: Properties and Physical Metallurgy.* Metals Park, OH: ASM International, 1984.

48   Hren JJ and Guy AG. *Elements of Physical Metallurgy.* Reading, MA: Addison Wesley, 1974.

49   *Metals Handbook, Vol. 4: Heat Treating.* 9th ed. Metals Park, OH: American Society for Metals, 1981.

50   *ASM Handbook, Vol. 20: Materials Selection and Design.* 1st ed. Metals Park, OH: ASM International, 1997.

51   Atlas Specialty Metals Tech Note No. 7, Galvanic Corrosion. Melbourne, Australia: Atlas Specialty Metal, May 2006.

52   American Society for Metals. Fabrication and Finishing. In: Horn KRV (ed.), *Aluminium*, Volume 3. Metals Park, OH: American Society for Metals, 1967.

53   Madhusudhan RG, Mastanaiah P, Murthy CVS, et al. *Microstructure, Residual Stress Distribution and Mechanical Properties of Friction-Stir AA 6061 Aluminium Alloy Weldments.* Indian Society for Non-Destructive Testing Hyderabad Chapter, Proc National Seminar on Non-Destructive evaluation Dec. 7 – 9, 2006, Hyderabad, Defence Research Development Laboratory, Hyderabad, India, December 2006.

54   Melhem GN, Munroe PR, Sorrell CC, et al. Field Trials of Aerospace Fasteners in Mechanical and Structural Applications. In: Totten GE, Tiryakioglu M, and Kessler O (eds.), *Encyclopedia of Aluminium and Its Alloys*, Vol. 1. Boca Raton, FL: CRC Press, 2019, pp. 1007–1021.

55   Concorde Structural Features: A description of the methods of construction and materials used in the production of the Anglo/French SST. *Aircraft Engineering and Aerospace Technology*, 43(4), 1971, pp. 16–22.

56   Harpur NF. Structural Development of the Concorde. *Aircraft Engineering and Aerospace Technology*, 40, 1968, pp. 18–30.
57   ASM Ready Reference: *Thermal Properties of Metals*. Materials Park, OH: ASM International, 2002, p. 11.
58   Thompson JG. *Nickel and Its Alloys*. Washington, D.C.: The National Bureau of Standards, 1958, p. 63.
59   Sequeira CAC. *High Temperature Corrosion: Fundamentals and Engineering*. Hoboken, New Jersey: Wiley, 2018, p. 24.
60   Orlebar C. *Concorde*. Oxford: Osprey Publishing, 2017, p. 138.
61   Froes FH. *Titanium Physical Metallurgy Processing Applications*. Materials Park, OH: ASM International, 2015, p. 362.

Supersonic Aircraft

2

Thermal Effects on Materials Used as Rivets

2.1 BACKGROUND ON THE CONCORDE

In the past, supersonic flight, and the attendant breaking of the sound barrier, were considered to represent next-generation commercial flight. Only two supersonic passenger planes have existed, the Anglo-French (British Aircraft Corporation [BAC] and French Aérospatiale). Concorde and the Soviet Tupolev Tu-144. The Concorde and Tupolev had wing spans of 25.6 m and 28.8 m respectively, and the length of the Concorde and Tupolev were 61.7 m and 65.5 m respectively. The Concorde fuselage expanded up to 300 mm while in supersonic flight due to aerodynamic heating, taking it up to about 62 m, then reducing in overall length again to 61.7 m after landing and cooling. The Tupolev was in service for only 10 years (31 December 1968–1 June 1978), but its service as a passenger aircraft endured only for seven months as it began flights on 1 November 1977. The Concorde was in service for 33 years (2 March 1969–24 October 2003), but its passenger service endured for 27 years, as it began on 21 January 1976. It cruised at an altitude of 60,000 ft or 18,300 m. Although work on the Concorde commenced earlier than on the Tupolev, it was fully constructed two months earlier than the Concorde. However, it has been rumoured by the British, that the Russians stole the Concorde designs from the French aerospace manufacturing company Aérospatiale in order to manufacture the Tupolev, and the British subsequently coined the Tupolev as

DOI:10.1201/9781003516903-2

Concordski [62]. The Tupolev had many design faults since it was based on outdated technology and expertise – and unlike the Concorde, it was very heavy and not designed to transport privileged classes of people.

The Anglo-French Concorde was famous for its transatlantic flights (between London, Paris, New York, and Washington), which would transport a maximum of 100 passengers to their destinations in record times approximately 50% of the norm, and due to time differences, arrive at times before they left [63]. The Concorde uses roughly 16 kg of fuel per 1 km of travel, therefore travelling from Paris to New York (just under 6,000 km), the aircraft carried and expended approximately 95 tonnes of fuel in one trip. There were only 14 Concorde aircraft that entered commercial service for British Airways and Air France (seven aircraft each), but in the year 1982, one of the Air France Concordes was retired to use its spare parts to service the other Concordes. One of the Concorde 203 aircraft that was operated by Air France was given a thorough inspection and maintenance on 21 July 2000 at Roissy Charles de Gaulle Airport in Paris. However, on 24 July 2000, British Airways discovered and reported that there were small hairline cracks on all of their seven Concorde aircraft spars or crossbeams of the wings. British Airways nonetheless insisted that these cracks did not pose any risk to the aircraft. On 25 July 2000, Air France Concorde crashed not long after it caught on fire after take-off from the Roissy Charles de Gaulle Airport, killing 100 passengers, nine crew members in the aircraft, and four people on land as a result (see **Figure 2.1**).

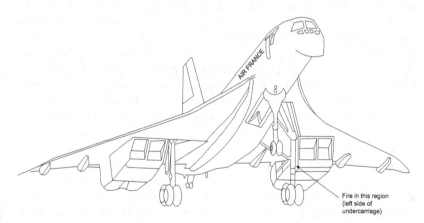

FIGURE 2.1 Concorde Left Tyre Ran Over a Titanium Strip, and it Caught on Fire Just After Take-Off from Roissy Charles de Gaulle Airport in Paris, and it Crashed onto a Hotel *La Patte d'Oie* in Gonesse, Killing 109 People on Board and Four in the Hotel.

One month after the crash of 25 July 2000, both Air France and Concorde stopped flights as this unfortunate incident highlighted the potential risks that these aircraft are highly vulnerable to. A thorough investigation into this crash suggests that the left tyre ran over a titanium metal strip which had rivet holes and some rivets in it. As the aeroplane took off and was close to 100 m in the air, the aircraft stalled with its nose pointing to the sky, and it rolled over to its left-hand side and went down and crashed. Despite earlier routine airport vehicle inspections along the runway, the strip was not located on the runway. This piece was a little under 500 mm long and about 20–30 mm wide, which was speculated by the investigators to have fallen off an earlier Continental Airlines DC 10 aircraft that was on the same runway 5 minutes prior. As it turns out, this piece of metal did match a similar section to the DC 10 right engine, lower left wear strip which was found to be missing from that particular DC 10 aircraft that had travelled along the same runway earlier. The titanium strip was meant to be manufactured from stainless steel material, and there were more holes in the strip than the original design had stipulated, which obviously demonstrates that the maintenance crew for the DC 10 used the wrong material and procedures to perform their repairs on the aircraft. The investigation also found that there had been 57 instances of ruptured tyres on the Concorde aircraft prior to this accident, and there had been instances where the fuel tanks were pierced as a result of ruptured tyres, causing projectiles to hit the tanks throughout the 27 years of flying, but no accidents or serious damage eventuated. As a result, it has been posited that since ruptured tyres can also cause serious damage to the tank by projectiles, excessive speed, angle of attack, new materials, and new technology, all remained esoteric. This meant that the variables involved in deciphering what happened, particularly when trying to compare or extrapolate the types of accidents involved with subsonic aircraft, was almost unintelligible even for the most experienced investigators. Within one year, after all of the investigations were made and improvements implemented in the Concorde aircraft, passenger flights by both airlines resumed once again. Two main improvements and preventative measures taken were: 1) fitting of the Michelin Near Zero Growth (NZG) tyres and 2) reinforcing most of the fuel tanks with composite Kevlar lining, which are typically used for bullet proof vests. However, the investigation also concluded with finding those that were responsible for the accident, such that Continental Airlines was fined for their negligence and they were convicted of manslaughter. One Continental Airlines mechanic was charged with a suspended prison sentence due to involuntary manslaughter, and was given a 15-month custodial sentence as he was found to be responsible for the incorrect installation of the metal wear strip in the DC 10 which led to the Concorde's fatal accident. In March 2003, the Concorde was finally suspended from service, and Air France ceased flying, while British Airways ceased flying in October 2003. Both Air France and British Airways stated that the reason for the cessation of flights was due to lack of passengers seeking supersonic travel. Almost every book or article written about the Concorde to date unanimously declares that while supersonic flight was shown to be technically feasible and successful, they were ultimately commercially unsuccessful.

2.2 BACKGROUND ON SUPERSONIC FLIGHT

When an aircraft travels below the speed of sound, it generates pressure waves, that is, sound waves, which are transmitted radially away from the aircraft and create minimal effect on the environment [64]. The speed of sound varies based on variables such as relative humidity. For example, at 30% relative humidity, the speed of sound ranges between 343.807–343.929 m/s, whereas at 60% relative humidity, the speed of sound ranges between 344.182–344.274 m/s [65]. As the relative humidity increases, the speed of sound also increases relatively. In contrast, when an aircraft reaches the speed of sound, the sound waves concentrate in front of the aircraft in the form of compressed air and create a shock wave that travels forward from the aircraft [66]. When an aircraft exceeds the speed of sound and hence becomes supersonic, it penetrates the compressed air volume and becomes subject to a significant drag barrier, which can cause the aircraft to vibrate intensely. Once it has passed through this volume, this effect is eliminated and the flight becomes stabilised. However, a supersonic aircraft continues to generate sound waves, which is directed towards the earth, thereby creating what are known as *sonic booms* along its flight path. These may be so intense that they can break windows.

Air speed is described in terms of the Mach number (M) [67] as per *Equation (2.1)*:

$$M = \frac{\text{Speed of Aircraft}}{\text{Speed of Sound}} \tag{2.1}$$

where: M < 1.0   Subsonic speed   (e.g., commercial aircraft)
        M = 1.0   Transonic speed
        1.0 < M ≤ 5.0   Supersonic speed   (e.g., fighter jets)
        M > 5.0   Hypersonic speed   (e.g., missiles)

One of the aircraft to inspire the development of supersonic flight was the Messerschmitt Me 163 Komet, which was the first rocket-propelled aircraft – it was also the first aircraft to exceed an air speed of 1,000 km/h in level flight [68]. This aircraft went into production as a fighter aircraft in early 1941, and it was able to cruise at a speed of 901– 949 km/h at its service ceiling altitude of 12,190 m [69]. At the speed of 1,062 km/h, the Me 163 Komet was able to travel at 90.3% the speed of sound (*viz.*, Mach 0.903) [70].

Two of the best-known supersonic aircraft are the commercial airliner Concorde and the reconnaissance aircraft Lockheed SR-71 Blackbird Military (otherwise known as SR-71 Blackbird). As mentioned earlier, the Concorde had its maiden flight in 1969 – but it entered service only in 1976 [62, 71], with only 20 built (these saw service until 2003, flying at Mach 2.04 between London/Paris and New York/Washington, D.C./Barbados) [60, 62]. The preliminary model of the SR-71 Blackbird first was flown in 1962 and the ultimate model first flew in 1964 and was capable of reaching a speed of Mach 3.3 [72]. Some of the current aircraft that travel at supersonic speeds include the Boeing F-15 Eagle, MiG-23, and Sukhoi Su-27, all of which are high-speed military aircraft that reach speeds of Mach 2.50, Mach 2.35, and Mach 2.35 respectively [73].

2.3 SELECTING MATERIALS FOR SUPERSONIC AEROSPACE APPLICATIONS

At these great speeds, a critical issue becomes heating from air friction, which is greatest at the leading surfaces of the aircraft [44]. These issues, which were recognised in the 1950s, have been addressed through developments in materials and their structural design. Consequently, creep resistant aluminium alloys were employed to improve resistance to deformation [74] and wings were reduced in size and orientation to reduce drag.

The overall flight characteristics and operating conditions of materials used in aircraft are relatively well understood from a macroscopic perspective, so the effects of heating on the leading edges of wings are well established [75]. Since the Concorde's fuselage expanded during supersonic flight, special paint was formulated and used on the aircraft in order that it will be able to withstand such an expansion in its overall length. However, there appear to be few data on the relevant micromechanical processes that took place within these materials that were in service. This is particularly the case with small components, such as the rivets joining the materials comprising the wings. Rivets represent one of the primary structural components of wings and other structural components of aircraft, but little is known of their localised responses to supersonic flight. As such, the present work examines aspects of the effects of supersonic flight on the thermal and mechanical stresses, responses to ageing, and corrosion of aerospace rivets. A case is made, in the last section of the book, to explain the requirement and possible methods in which to research into the various temperatures that occur along the rivet's shaft and mandrel below the aircraft skin, particularly while in supersonic travel where heating of the skin is most severe.

Unlike traditional subsonic aircraft, the components on the Concorde required tailored designs and manufacturing processes in order to withstand the extreme conditions during supersonic flight. Standard aluminium alloys, such as AA7075-T6 and AA2024-T4, which are used commonly for fasteners such as rivets, exhibit gradual strength decreases relative to those at ambient upon heating and, at the extreme, are limited to temperatures of 150–200°C [75, 76]. In fact, aircraft experience significant material softening in certain areas when supersonic speeds are encountered, whereas in subsonic aircraft, the temperatures in the same regions do not reach in excess of 80–90°C [36]. These temperatures are well below the standard maximal operating temperature of 127°C for the nose and leading edges of the superstructure of the Concorde in flight [77]. Consequently, the fabrication of a thermally stable superstructure or skin required the engineering of Hiduminium RR58 aluminium alloy (Al-2.5Cu-1.5Mg-1.2Ni-1.0Fe-0.2Si-0.1Ti), which originally had been developed during WWII [78]; this alloy also is known as Al alloys AA2618-T6 [78] and CM001 [79]. CM001 has been altered to improve its creep resistance, and the initials CM are derived from *Concorde Material*. That there are several names for this alloy is due to a few countries such as the English, French, and Americans designating the alloy – for instance RR58 (English designation), Hiduminium (or Hi-Du-Minium short for High Duty Aluminium by Rolls Royce) also known as AA2618 or 2618-T6 (American designation), and AU2GN (French designation). Overall, when comparing all the aluminium alloys, AA2618-T6 or CM001 have the best strength and fatigue properties when exposed to a large range of temperature. More refractory alloys in the form of Ni superalloys and lightweight Ti alloys have been used for more thermally sensitive areas, particularly around the engine. Again, there appear to be few data concerning the effects of supersonic flight on these materials; however, the present work is restricted to the effects of thermal and mechanical stresses, ageing, and corrosion on such materials when used as rivets.

The titanium rivets of interest in this book are manufactured from a titanium alloy, Ti-6Al-4V, which is considered to be the workhorse alloy of the titanium industry and comprises more than 50% of the titanium alloys used worldwide. Titanium alloys are used widely in aerospace structures due principally to their high strength and creep resistance and excellent corrosion and oxidation resistance. Titanium alloys, in general, retain their strength at elevated temperatures and are therefore used in the vicinity of the engine, as they provide better heat resistance (and offer considerable weight reduction when compared to steel).

Elemental titanium undergoes an allotropic phase transformation from a hexagonal close-packed (HCP) α phase, stable at room temperature, to a body-centred cubic (BCC) β phase at approximately 882°C. Alloying additions to titanium can strongly affect the α to β transition temperature, called the β-transus, the shape of the α + β phase field, the nature of any transformations between these phases during heat treatment and hence the microstructure and properties of these alloys. Alloying additions, such as aluminium and oxygen, increase the β-transus and strengthen the α phase, and are termed α stabilisers. Conversely, alloying elements, such as vanadium and molybdenum, decrease the β-transus and strengthen the β phase and are termed β stabilisers. The HCP α phase endows titanium alloys with good creep resistance, but the presence of this phase can lower ductility. On the other hand, the presence of the β phase enhances ductility and formability. Titanium alloys are classified by their principal alloying additions, and Ti-6Al-4V is termed an α + β alloy [80]. It exhibits an α + β phase field that spans a wide temperature range and so can be readily heat treated to exhibit both allotropic phases and a range of structures combining both phases in different morphologies. As such, careful control of heat treatment of Ti-6Al-4V can be used to tailor the microstructure of this alloy to generate a wide range of possible mechanical properties. For example, slow cooling from the β phase can lead to a microstructure of coarse α laths that promote creep resistance, whereas rapid cooling from the α + β phase field can lead to refined microstructures that exhibit high strength. Alloys that are heavily alloyed with β stabilisers, termed β alloys, can be heat treated to a microstructure of β matrix containing α precipitates that can generate very high strengths. For example, Ti-15Mo-5Zr-3Al (where the α stabiliser, Al, is added to promote α precipitation) can exhibit room temperature yield strengths of nearly 1,500 MPa. Rivets manufactured from titanium alloys are commonly either α + β alloys or β alloys.

2.4 RIVET LOCATIONS ON SUBSONIC AND SUPERSONIC AIRCRAFT

Rivets are used on aircraft wings in the spanwise and chordwise direction to connect the internal structures of the wing to the skin. In the spanwise direction, spars and stringers are connected to the skin and in the chordwise direction, ribs and nose ribs are connected to the skin. **Figure 2.2** is a SolidWorks drawing of a rectangular wing, which clearly shows the internal structure of the wing. The main structural components, such as the ribs and spars, are clearly shown with their respective rivet locations. Along the leading edge and trailing edge of the wing, the front wing spar and rear wing spar are connected to the

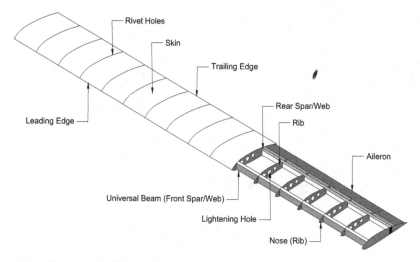

FIGURE 2.2 A SolidWorks Drawing of a Wing, Showing the Internal Structure.

aircraft's skin by rivets, respectively. Supersonic and subsonic aircraft use the same layout shown in **Figure 2.2** for different wing shapes and profiles. A very similar internal structure is used for the empennage structure, which consists of the horizontal and vertical stabiliser.

Furthermore, rivets are also used to attach the stringers, longerons and formers to the skin of the fuselage. The internal structure of a semi-monocoque fuselage panel is shown in **Figure 2.3**, clearly showing its structural components. Most aluminium rivets in general aircraft that are used to connect internal structures to the skin require a transition or interference fit, since minimum rivet movement is required when the aircraft is under load during flight or on the ground. An interference fit is preferred since it allows the material around the rivet hole to be compressed which further results in exceptional load transfer. Hence, positioning of rivets on supersonic aircraft is dependent on the positioning of the internal structure of the wing, fuselage, and other structural components.

2.5 SKIN FRICTION AND DRAG EFFECTS IN SUPERSONIC AIRCRAFT

As the aircraft travels through the air, there is movement of air molecules adjacent to the aircraft. Depending upon the speed of the aircraft, the air molecules

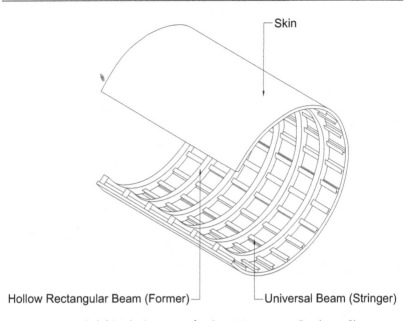

Hollow Rectangular Beam (Former)┘   └Universal Beam (Stringer)

FIGURE 2.3   A SolidWorks Drawing of a Semi-Monocoque Fuselage, Showing its Internal Structure.

are affected in varying ways. Subsonic aircraft are designed for use at temperatures between −55 and 80°C. Speeds below Mach 1 do not change the density of the air at all and the flow is incompressible. At greater speeds, when Mach is greater than 1, the flow is supersonic, and the air density is changed and displaced, making compression pronounced, which affects the aircraft. As the speed of the aircraft increases beyond the speed of sound (340 ms^{-1}), shock waves are created that affect both drag and lift on an aircraft. A cone of pressurised molecules is a result of the shock wave, and spreads outwards and rearwards and extends down to the ground. This is heard similarly to thunder from lightning from the ground when the aircraft passes directly overhead and is referred to as the sonic boom (see **Figure 2.4**).

Certainly, airframe temperatures fluctuate on a larger scale in supersonic aircraft as opposed to subsonic aircraft due primarily to skin friction effects. If the speed is at exactly Mach 1 and equal to the speed of sound or when the aircraft has parts exposed to flow that is subsonic and other parts to supersonic, then this is termed transonic. Due to the decreasing ambient air temperatures as altitude increases, the temperature on the skin of an aircraft firstly decreases. Also, it has been speculated that since there is less oxygen potential at higher altitudes and a lesser atmospheric pressure than we experience on the ground,

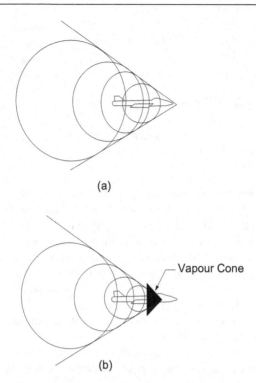

(a)

(b)

Vapour Cone

FIGURE 2.4 (a) F/A – 18 Hornet Breaking the Sound Barrier Creating a Sonic Boom; (b) F/A – 18 Producing Vapour Cloud Due to Water in Air Condensing.

there is a corresponding lowering of oxidation effects. Frictional affects can create high temperatures on the leading edge of aircraft when travelling above Mach 1. At a speed of Mach 2, the temperature on the skin remains at 100°C (212°F). At a speed of Mach 2.2, the temperature starts to rise and the temperature on the skin reaches up to 120°C (248°F) [44]. At a speed of Mach 2.4 the temperature on the skin reaches up to 150°C (302°F). During descent, the reverse occurs [44].

Interestingly, at a speed of Mach 2, the Concorde heats up to a diverse range of temperatures in different regions of the structure (see **Figure 2.5**). For instance, a temperature of 114.5°C up to 127°C is reached at both the nose and trailing edge and averages around 95–98°C at the tail, and the temperatures on the skin can reach up to 149°C at this speed. The Concorde can also stretch up to as much as 152.4–254 mm in length [81] due to the heating of the airframe when it is cruising at a speed of Mach 2. At a speed of Mach 5, the aircraft is hypersonic.

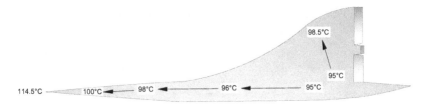

FIGURE 2.5   Schematic Representation of the Heating Associated with the Concorde Cruising at Mach 2.04.

Although aerodynamic engineering is much more detailed and more complex than outlined here, it is worth mentioning in brief that one of the main properties studied in this field is what is termed the concept of the boundary layer. The relevance of this is the effect of temperature, and it is these temperatures that must be taken into account for every single component in an aircraft. The boundary layer plays a major role in determining the drag of an aircraft, and it is drag that creates skin friction. Therefore, the wings and other components should be designed to reduce drag. Drag affects the entire surface of an aircraft and is not limited to any region, however, the severity of drag depends upon the contours and surface roughness also.

If we are to consider a basic aerofoil as shown in the schematic below in **Figure 2.6**, then we can see that air passes over the wing on the leading edge and travels towards the centre of the wing. Initially, the flow is smooth and uninterrupted over the wings, fuselage, and other parts of the aircraft and this is known as the laminar layer. As the wing continues moving through the air, it tends to reduce in speed due to the creation of skin friction, thus making the layer thicker and more turbulent and this layer is known as the turbulent layer. From this point on, the turbulence continues and there is an accompanied loss of lift and increased drag. The transition point is the junction where the laminar layer changes to a turbulent layer. There are various factors that affect the transition such as surface temperature, surface roughness, and speed. The transition point therefore moves forward as the speed increases and the angle of attack increases (see **Figure 2.7**). The turbulent layer is thicker than the laminar layer, so it produces higher skin friction. In turn, this skin friction drag causes another type of drag called pressure drag. Therefore, if we design for either an increase or decrease in both pressure drag and friction drag, we must design for either thin body for skin friction or blunt (cylindrical or spherical shapes) bodies for pressure drag.

Golf balls have dimples rather than being smooth to create a turbulent boundary layer at the front so that the drag is reduced from behind while it is in flight. This is often referred to as "tripping" the boundary layer. These

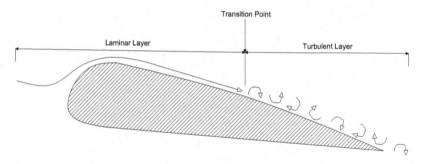

FIGURE 2.6 Boundary Layer Being Converted or Transitioning into Laminar and Turbulent Layer.

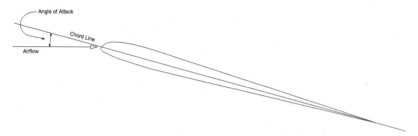

FIGURE 2.7 Lift Varies with the Angle of Attack (the Greater the Angle, the Greater the Lift).

dimples are only suitable for designs on cylindrical or spherical shapes. However, there are aircraft such as the vortex that take off and land at low speeds that utilise something similar to the concept of dimples on their wings in order to create turbulence, delaying wing stall, thereby encouraging an increase in lift. The vortex generators basically create vortices of air across the wings or other surfaces of the aircraft, and they regulate the divergence of airflow, which increases the lift and performance of the aircraft. These vortex generators typically comprise small rectangular plates that are positioned as shown below in **Figure 2.8** on the wing along three rows. There are many other methods that may be used, and each has its advantages and disadvantages – but these options are only necessary when redesigning the wing is not an option.

Of main concern is the degradation of material strength at elevated temperatures, particularly with higher speeds. One area of interest is the ageing of materials, since metals upon elevated temperatures generally lose strength, and this reduction in strength is dependent upon the type of metal. Frictional heating for aircraft travelling at greater than Mach 2.5 means that lightweight

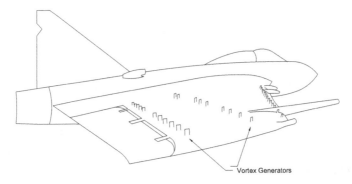

FIGURE 2.8 A Gloster Javelin Showing the Rectangular Vortex Generators Aid Higher Speed Flow Over the Wings to Delay Wing Stall.

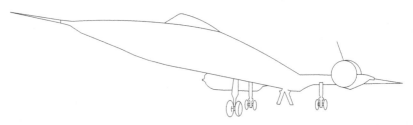

FIGURE 2.9 The Central Intelligence Agency (CIA) Was Involved with Ensuring a Sweeping Design for the Lockheed SR-71 Blackbird Military, in Order to Counteract Soviet Radar. It Reaches an Impressive Velocity of up to Mach 3.5 at Altitudes of 21,300–27,400 m.

aluminium is not capable of withstanding the temperatures generated. For instance, titanium is light enough and has good high temperature resistance and was used in approximately 90% of the SR-71 Blackbird (see **Figure 2.9**) and the entire skin comprised of titanium alloy, while the remaining materials were highly developed composite materials.

It was the excessive heating of the leading edge of the left wing in the Space Shuttle Columbia in 2003 during its re-entry into the earth's atmosphere that caused the catastrophe that killed seven astronauts. However, the breach was due to a large piece of insulating foam from the external tank being dislodged and hit the leading edge of the space shuttle's left wing, creating a hole in the Reinforced Carbon-Carbon (RCC) panel on the leading edge of the wing on its underside during take-off. Upon re-entry, the extra heat from the superheated gases penetrated the wing and led to its ultimate failure. However, although the temperatures imposed upon materials are very severe during re-entry, this highlights the unfortunate reality of material design, which is that every material has thermal properties up to a certain point before final degradation.

2.6 ESTIMATING TEMPERATURE ON RIVETS IN SUPERSONIC AIRCRAFT

Atmospheric data under the US Standard Atmosphere and The International Standard Atmosphere can be used in conjunction with *Equation (2.2)* to estimate the temperature of rivets operating under supersonic conditions. Aircraft Skin Temperature Equation [7], skin friction and drag effects in aircraft:

$$T_s = T_a \left(1 + 0.2\, M^2 \right) \tag{2.2}$$

where: T_s = Stagnation Temperature (K)
$\quad\;\; T_a$ = Ambient Temperature (K)
$\quad\;\; M$ = Mach Number

The basic thermodynamic theory described in *Equation (2.2)* was essentially derived from the expression for the isentropic process. *Equation (2.2)* provides a temperature estimate of the skin temperature over the airframe of the aircraft at a required operating condition. The Concorde's cruise altitude was between 12,000–18,000 m [82], and according to US Standard Atmosphere, the aircraft would be exposed to an atmospheric temperature of –56.5°C (216.65K) [83]. By applying *Equation (2.2)*, with the Concorde's Mach number 2.04, a stagnation temperature of 123.82°C (396.97K) is yielded and this approximation provides an error of 2.50% when compared to an average thermocouple reading of 127°C, from an in-flight Concorde flight [79]. From this divergence, it is possible to postulate reasons for these differences: 1) instrumentation inaccuracy, where there may be a slight error in the readings of the thermocouples; 2) varying Mach number, that is, Mach 2.2 (this was Concorde's maximum Mach number speed [11] which provides a temperature of 153.22°C (426.37K)); 3) surface finish affecting flow and compressibility behaviour; 4) slight discrepancy due to the atmospheric (static) temperature at altitude being slightly higher (greater than -56.5°C); or 5) a culmination of all the preceding factors mentioned above. **Figure 2.10** uses *Equation (2.2)*

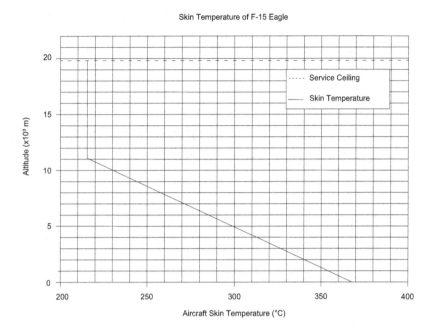

FIGURE 2.10 A Graph Showing the Skin Temperature of the F-15 Eagle at Various Altitudes for its Cruise Mach Number of 2.5.

to provide the stagnation aircraft skin temperature for the F-15 Eagle at its cruise Mach number of 2.5, at various altitudes. This graph clearly shows one of many reasons that supersonic fighter aircraft fly above 10,973 m. This is because, as the altitude decreases, the skin temperature linearly increases as shown in **Figure 2.10**.

Another method in estimating the average skin temperature of an aircraft is to obtain a temperature contour of the aircraft, as shown in **Figure 2.5** above, and to add the temperatures on the aircraft's skin (excluding the leading edge) and find the average temperature. For example, from **Figure 2.5** the average skin temperature of the Concorde is 99.57°C, and from **Figure 2.11** the skin temperature value obtained is 106°C, which is found by using a speed of Mach 2.04 [60] at an altitude of 15,635 m. As a result, the average skin temperature of the Concorde is estimated at 102.79°C.

In addition, it was acknowledged by Raymer that *Equation (2.2)* had limitations since "actual skin temperatures are difficult to calculate because they depend on airflow conditions, surface finish, and atmospheric conditions" [7]. Other factors include heat radiating from engines and

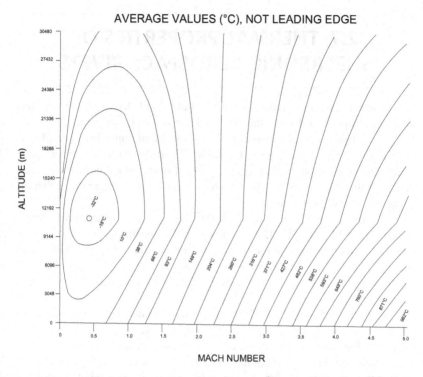

FIGURE 2.11   Mach Number vs Altitude Contour Plot for Aircraft Skin Temperature [7].

the aircraft. Furthermore, as time t → ∞ (that is, where "t" represents the flight duration), the wall temperature of a diathermic aerofoil (an aerofoil in which heat transfer occurs) will be uniformly distributed at the value of the stagnation temperature given by the thermodynamic equation and this characteristic is known as thermal equilibrium. The surface temperature of a diathermic object becomes more uniform as the thermal conductivity of the model increases.

In other words, the surface of an object that has high thermal conductivity will attain thermal equilibrium faster than an object that has low thermal conductivity. The limitation behind this model is that practically no supersonic aircraft operates for long periods and therefore will never achieve thermal equilibrium. However, the methods that are utilised to calculate the skin temperatures can serve as a useful heuristic in estimating the surface temperature of the rivet.

2.7 THERMAL PROPERTIES OF SUPERSONIC AEROSPACE RIVETS

Aerospace rivets are predominately manufactured from aluminium and titanium alloys, each having their own unique material properties. The coefficient of linear thermal expansion (CTE) for aluminium alloys is high and is in the range of 16–24 μm/m°C. **Table 2.1** shows the CTEs of common materials that are used for the manufacturing of rivets, when they are subjected to temperatures between 20–200°C, which was the temperature range that the Concorde was subjected to.

For comparative purposes, it is interesting to note that steel, for instance, has a CTE of 11 μm/m°C [85]. As a result, it is clear that Ti-6Al-4V does not expand as much as the aluminium alloys, hence, it has a higher resistance to aerodynamic heating. Furthermore, if there is a mismatch of material between the rivet and the skin, residual stress may occur due to the difference in thermal expansion coefficients.

While the difference in CTE within each of the materials (not comparing to other materials) may seem almost negligible, there is little to no information in literature regarding the constant expansion and contraction which affects the overall fatigue life of the metal as a function of flight cycles. During flight, approximately 200 mm expansion in the overall structure of Concorde's fuselage has been reported [86], indicating significant heating effects on AA2618-T6 (CM001) during flight. The 1968 study by Harpur [87] indicates that at cruising speeds, there is significant thermal energy which may cause sustained creep strain. If the effects of thermally induced strain are considered, it is then possible to plastically deform the material such that the overall strain fields are relieved, adding to the degradation of the skin material. While this may appear insignificant over a few cycles, when the entire lifespan of a Concorde is considered, this can cause significant wear in the overall structure.

TABLE 2.1   CTEs of Aluminium and Titanium Alloys

MATERIAL	THERMAL EXPANSION (20°C - 200°C)	REFERENCE
AA2618-T6	22.9 μm/m°C	[84]
AA7075-T6	24.3 μm/m°C	[84]
Ti-6Al-4V	9.40 μm/m°C	[84]

2.8 FATIGUE TESTING OF AEROSPACE RIVETS ON SUPERSONIC AIRCRAFT

The then British Airways Chairman David Nicholson in 1974 stated: "We don't know what the working life of the machine [Concorde] will be" [87]. This was stated due to the uncertainties that the Concorde provided, and these uncertainties were attributed mainly due to aerodynamic heating which was caused by the friction between the Concorde's skin and the airflow in contact with the skin at high Mach numbers. Hence, the temperature of the skin increases due to friction, and this further implies higher thermal stresses [87] on the structural components of the Concorde, which includes the rivets. The rivets protrude through the skin, and hence aerodynamic heating is present on the surface (head) of the rivets. Airflow does not come into direct contact with the shank of the rivet as the skin provides a barrier. This was evident during experimental testing on the Concorde, which clearly highlighted the temperature difference between the skin and the mid spar/web structure inside the wing [88] as shown in **Figure 2.12**.

The mid spar/web is not exposed to the airflow [7], but the temperature continues to increase due to the conduction of heat from the skin. **Figure 2.12** clearly shows this phenomenon of how the skin is connected to the mid spar/ web. The rivet's head is subjected to the airflow, and the shank is not, since it is embedded in a region in the vicinity of the mid spar/web. The use of flush rivets, also known as countersunk rivets, ensures that the total drag accumulation is at a minimum, and as a result will be much lower when compared to button head or round head rivets [6]. In other words, it provides *aerodynamic cleanness*. Another reason for using flush rivets instead of round/button rivets was that only the top surface of the flush rivet, which was a very small surface area, will encounter aerodynamic heating and hence high temperatures.

In relation to the structural integrity of supersonic aircraft, that is, the Concorde, it is evident in literature that it has been difficult to conduct testing on a supersonic aircraft's fatigue life, due to the existence of aerodynamic heating which leads to high thermal stresses, causing "expansion", "overageing", and "creep". The main problem for fatigue testing occurs not due to the reasons mentioned above, but due to the fact that they are time-dependent. Hence, the skin, and as a result the rivets, are exposed to high temperatures for an extended period of time. Supersonic transport (SST) fatigue tests on the Concorde were set up to ensure that the Concorde being tested was able to withstand three times as much fatigue damage [56] than the verified Concorde in service. As a result, the fatigue tests were accelerated to achieve this goal.

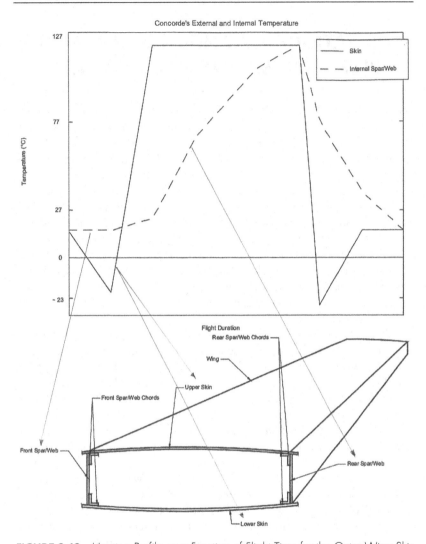

FIGURE 2.12 Heating Profile as a Function of Flight Time for the Outer Wing Skin and Inner Spar/Web Showing a Distinct Temperature Gradient Between the Skin and Inner Spar/Web [88]. The Schematic of the Wing Structure is Analogous to a Structural I-Beam Where the Flanges Would Represent the Skin on the Aircraft and the Web Represents the Spar/Web on the Aircraft.

Fatigue testing on the Concorde was conducted by using three cycles (two hot test cycles and one cold test cycle). The hot cycles consisted of mechanical loading and an accelerated thermal cycle, whereas the cold cycle consisted of

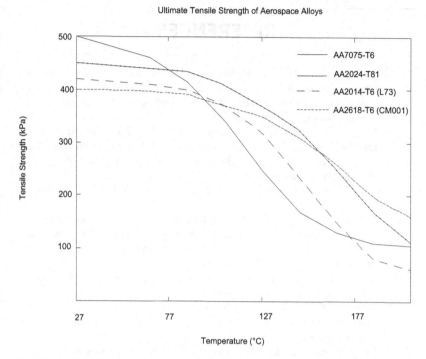

Ultimate Tensile Strength of Aerospace Alloys

FIGURE 2.13   Ultimate Tensile Strengths of Various Aerospace Alloys After 20,000 Hours at Elevated Temperatures [56].

mechanical cyclic loading experienced by subsonic aircraft. In relation to the hot cycle, the Concorde's fatigue testing was accelerated by subjecting the structure to a temperature of 130°C by applying "hot soaking" or by using infrared lamps [56]. An example of a soak test result on the Concorde is shown in **Figure 2.13**, and this shows a comparison between AA2618-T6 (CM001), AA2024-T81, AA7075-T6, and AA2014-T6 (L73) (Duralumin), when these materials have been subjected to 20,000+ hours of testing, while being exposed to the fight conditions of the Concorde. It was evident that AA2618-T6 (CM001) can withstand higher temperatures and maintain its tensile strength when compared to other materials.

In relation to the cold cycle, the mechanical cyclic loading consisted of taxiing loads, take-off loads, gust loads, fuel handling, and cabin pressurisation [56]. Mechanical cyclic loading on the Concorde occurred when the aircraft temperature was low, which occurs during climb and descent. Combining both cycles results in a major fatigue test conducted on the Concorde. In relation to the structural integrity of the rivets, these rivets connect the internal structures to the skin, hence the rivets are also subjected to these conditions as they are structural components of the aircraft.

REFERENCES

6   Wright B, Dyer WE and Martin R. *Flight Construction and Maintenance.* Chicago: American Technical Society, 1941, pp. 137–138.

7   Raymer DP. *Aircraft Design: A Conceptual Approach.* 6th ed. Reston, Virginia: AIAA, 2018, pp. 514–520.

11   Melhem GN. Aerospace Fasteners: Use in Structural Applications. Encyclopedia of Aluminium and Its Alloys. In: Totten GE, Tiryakioglu M and Kessler O (eds), *Encyclopedia of Aluminium and Its Alloys*, Vol. 1. Boca Raton, FL: CRC Press, 2019, pp. 30–45.

36   Mouritz A. *Introduction to Aerospace Materials.* Oxford: Woodhead Publishing, 2012.

44   Committee on Evaluation of Long-Term Aging of Materials and Structures Using Accelerated Test Methods. *Accelerated Aging of Materials and Structures, the Effects of Long-Term Elevated-Temperature Exposure.* Washington, DC: National Academy Press, 1996.

56   Harpur NF. Structural Development of the Concorde. *Aircraft Engineering and Aerospace Technology*, 40, 1968, pp.18–25.

60   Orlebar C. *Concorde.* Oxford, UK: Osprey, 2017, p. 138.

62   Morris N. *Transport (From FAIL to WIN! Learning From Bad ideas).* UK ed. Raintree, 2011.

63   Curley R. *The Complete History of Aviation: From Ballooning to Supersonic Flight.* Britannica Educational Publishing, 2012, p. 102.

64   US Department of Transportation. *Transportation Noise and Its Control.* Washington, DC: U.S. Government Printing Office, 1972, p. 4.

65   Evans LG and Bass HE. *Tables of Absorption and Velocity of Sound in Still Air at 68°F (20°C).* Report WR 72-2. Huntsville, AL: Wyle Laboratories, January 1972.

66   Abbott IH and Doenhoff AEV. *Theory of Wing Sections.* 2nd ed. London: Constable and Company, 1959, p. 248.

67   National Aeronautics and Space Administration. *NASA Technical Translation*, Issue 290. Washington, DC: NASA, 1965, p. 8.

68   Schmidt E. *Black Tulip: The Life and Myth of Erich Hartmann, the World's Top Fighter Ace.* Havertown, PA: Casemate Publishers, 2020, p. 88.

69   United States Air Force Museum. Illustrated History. Dayton, OH: Air Force Museum Foundation. 1987, pp. 56.

70   Kuethe AN and Chow C. *Foundations of Aerodynamics.* 4th ed. Toronto, ON: Canada: Wiley, 1986, pp. 203–204.

71   Linden FRVD, Spencer AM, and Paone TJ. *Milestones of Flight: The Epic of Aviation with the National Air and Space Museum.* New York: Quarto, 2016, p. 192.

72   Graham RH. *SR-71 Blackbird: Stories, Tales, and Legends.* Zenith Imprint, 2002, p. 223.

73   Taylor JWR. *Jane's All World Aircraft 1985–86.* London: Jane's Publishing, 1985, pp. 237, 254, 451.

74   Robinson JS, Cudd RL, and Evans JMB. Creep Resistant Aluminium Alloys and Their Applications. *Materials Science and Technology*, 19(2), 2003, pp. 143–155.

75   Kaufman JG. *Fire Resistance of Aluminium and Aluminium Alloys and Measuring the Effects of Fire Exposure on the Properties of Aluminium Alloys.* Materials Park, OH: ASM International, 2016.

76   Leyman CS. A Review of the Technical Development of Concorde. *Progress in Aerospace Sciences*, 23, 1986, pp. 185–238.

77   Doyle WM. The Development of Hiduminium-RR.58 Aluminium Alloy: The Background to the Choice of the Main Structural Material for Concorde. *Aircraft Engineering and Aerospace Technology*, 41(11), 1969, pp. 11–14.

78   Hiduminium RR. *Aerospace Structural Metals Handbook*, Vol. 3. West Lafayette, IN: Purdue University, 1989.

79   Royster DM. *Tensile Properties and Creep Strength of Three Aluminum Alloys Exposed up to 25 000 Hours at 200^0 to 400^0 F (370^0 to 480^0 K).* NASA TN D-5010. Washington, D.C.: National Aeronautics and Space Administration, 1969, p. 2.

80   Polmear IJ. *Light Alloys: From Traditional Alloys to Nanocrystals.* 4th ed. Oxford: Butterworth-Heinemann, 2005, pp. 299–335.

81   Cefrey H. *Super Jumbo Jets Inside and Out.* 1st ed. New York: PowerPlus Books, 2002, p. 14.

82   Cramoisi G. *Air Crash Investigations: The End of the Concorde Era, The Crash of Air France Flight 4590.* Morrisville, NC: Mabuhay Publishing, 2011, p. 506.

83   United States Navy Hydrographic Office. *Air Navigation.* Washington, D.C.: US Government Printing Office, 1963, p. 75.

84   Nygren WD. Development of a High Force Thermal Latch. *29th Aerospace Mechanisms Symposium hosted at South Shore Harbour Conference Facility, League City, TX*, 1995, p. 164.

85   Coyle JA. *Aviation Electrician's Mate 3 & 2.* Washington, DC: Naval Education and Training Program Development Center, 1981, pp. 2–34.

86   Linden FRVD. *Best of the National Air and Space Museum.* 1st ed. Washington, DC: Smithsonian, 2006, p. 41.

87   Harpur NF. Concorde Structural Development. *Journal of Aircraft*, 5(2), 1968, p. 180.

88   Peel, C.J. Degradation of Aluminum Alloys—Concorde Experience. Paper presented at the Workshop on Long-Term Aging of Materials and Structures, National Materials Advisory Board. Washington, D.C.: National Research Council, August 10–12, 1994.

Active and Future Research

3

3.1 RATIONALE FOR DESIGN OF RIVETS IN SUPERSONIC FLIGHT

The subsonic flight time by traditional travel between London and New York, for instance, would take approximately seven hours, but the same flight path was reduced due to supersonic travel to just under three hours. Travelling even longer distances by subsonic aircraft, for instance between Los Angeles to Sydney, would take over 17 hours, but by supersonic flight the time would be reduced to approximately nine hours. Myriad variables need to be considered in the design to achieve a reduction of close to 50% in flight time, one of which necessitated a design that would impart a high degree of efficiency, and two of the most important of these considerations that can be briefly discussed are the engines and materials that were used in the Concorde. First, engines to be considered for the Concorde had to achieve high-speed, but also allow for a smooth transition of air intake into the engines without the shock waves creating any damage to any parts of the aircraft. The energy required to achieve the speed of Mach 2.0 was not without a trade-off, as the combustion effects of the fuel were not entirely safe nor environmentally friendly, and the noise levels were difficult to reduce further without reduction in aircraft thrust. Second, the materials used on the Concorde were predominantly selected based upon the speed of the aircraft, since heating is caused as a result of the friction created by the contact of air with the aircraft's skin. Of particular note is that the fuel was used to cool the skin during flight. However, what was most reviled by many aerospace experts at the time of design was that the engines were too close to the fuel tanks which posed severe hazards. For instance, fuel could potentially flow from the fuel tanks into the engine bay, allowing ignition sources within the engine to create fire and explosion.

Engines that were originally considered in the design for the Concorde were Turbofans (Olympus 593), however, due to the large size of the turbofans, they were deemed to create excessive drag effect. On this basis alone, the design

DOI:10.1201/9781003516903-3

incorporating turbofans was eliminated, and turbojets instead were placed in the Concorde. Since traditional aircraft engines are only able to process air around Mach 0.5, the air intake for the Concorde travelling at Mach 2 must be reduced by decelerating the air intake down from Mach 2 to Mach 0.5; otherwise, if the air intake is not slowed down, shock waves would be generated which ultimately will cause damage to the engines.

Materials on the Concorde were selected based upon the temperature generated due to the speed of the aircraft. The speed of Mach 2 was chosen, since an estimated life of 50,000 hours (10 to 12 years flight time) was typically required of the Concorde when exposed to maximum temperatures ranging between 120–180°C. The speed of Mach 2 was therefore the limit for supersonic aircraft aluminium alloys, as the alloy would be significantly weakened above this temperature range. If a higher speed than Mach 2 is required, then other materials such as titanium or stainless steel must be considered in lieu of Hiduminium aluminium alloy AA2618-T6 (CM001). During subsonic flights, an aircraft experiences lower temperatures on its skin at higher altitudes in the troposphere (maximum flight altitude is 40,000 ft or 12,200 m), whereas in an aircraft travelling at supersonic speed, the skin of the aircraft increases rapidly due to certain aerodynamic factors involving stagnating flow (flow of air stagnates) which converts kinetic energy into heat (maximum flight height is 60,000 ft or 18,300 m). Typical regions involving stagnation are the nose, leading edges of the wings, engine inlets, and tails. Generally, the higher the altitude, the lower the temperature. The temperature of the Concorde drops while rising in altitude, but upon approaching a speed around Mach 1, the aircraft skin temperature begins to increase. Upon attaining a speed of Mach 2, the skin temperature reaches about 120°C, and at Mach 2.4, the skin temperature would reach about 150°C. While the aircraft descends, the skin temperatures reverse. If the Concorde were to travel at the lower altitudes similar to subsonic aircraft, but at a speed of Mach 2 (just over 2 times faster than the speed of a subsonic aircraft), then this would, again, involve even higher drag and higher heating temperatures on the skin of the aircraft. The design intent for incorporation of the Hiduminium aluminium alloy in the Concorde, as presented earlier, is to withstand higher temperatures and not compromise the original tensile strength when compared to many other materials. This aluminium alloy increases in strength without any sacrifice in ductility at ambient (20°C), and even when it is subjected to elevated temperatures up to 300°C. The copper and magnesium in this alloy improve the strength, whereas the nickel and iron improve hardening and creep resistance. It must be noted, however, that the trade-off for aluminium alloys is that the alloying element copper tends to decrease the overall corrosion resistance of the base aluminium alloy more than any other alloying element, and has the potential for intergranular corrosion, stress corrosion cracking and pitting.

Since different regions are exposed to differing temperatures, then these different regions are subjected to differing thermal variations and material expansion. For instance, AA7075-T6 or AA2618-T6 (CM001) are both aluminium alloys but with differing chemical compositions – and as shown in **Chapter 2**, **Figure 2.13**, AA7075-T6 materials at 127°C maintain a tensile stress of 250 kPa after 20,000 hours, whereas AA2618-T6 (CM001) have an even higher tensile stress of 360 kPa at the same temperature and same duration time of 20,000 hours. Therefore, AA7075-T6 reaches a tensile strength that is 70% the tensile strength of AA2618-T6 (CM001) in the same environment at 127°C and duration of time, 20,000 hours. In the range of 20–200°C, AA2618-T6 (CM001) has a CTE which is 2 μm/m°C less than the CTE of AA7075-T6 (see **Table 2.1** in **Chapter 2**). Due to the large variety of materials on the aircraft, creep strain is likely when materials with higher CTE differences are present and fatigue cracks eventually develop within these materials over a long period of time (and these small cracks may not be noticeable in the earlier stages). The basic differences in various materials in an aircraft indicating varying thermal variations and material expansion is therefore evident from this simple extrapolation. Rivets, for instance, would be made of a different alloy material to the Concorde skin – and since the skin is mostly AA2618-T6 (CM001) or RR58, the rivets would be manufactured from titanium alloy since it is light and has good resistance to high temperature. It is clear that supersonic aircraft would benefit from the use of titanium alloys as demonstrated in the SR-71 Blackbird, as titanium has been used successfully in 90% of the Blackbird's skin. For comparative purposes, titanium alloys (Ti-6Al-4V) which would have a CTE of 9.40 μm/m°C, AA2618-T6 (CM001) has a CTE of 22.9 μm/m°C, and AA7075-T6 has a CTE of 24.3 μm/m°C [84]. The CTE of titanium is almost 2/5th that of the CTE of the aluminium alloys, so this is a large difference in expansion between any of these two materials. Since titanium alloys have a much lower CTE to aluminium alloys, the relative location of the titanium material, for instance the rivet hole positioning, would remain precisely where it was originally made, even at varying temperatures. The negative aspect of titanium is that it is very difficult to machine as compared to that of aluminium alloys (and is more costly). **Chapter 2**, **Figure 2.12**, shows that the rivet shank is relatively unaffected by airflow as the rivet head and skin face are tightly sealed from the airflow, and this was evident during a Concorde experimental test that demonstrated there was a temperature difference between the mid spar/web and the skin inside the wing [88]. Since the rivet extends beneath the skin, it is akin to the mid spar/web being in contact with the skin – but also beneath it.

However, there is scarce information in literature regarding rivets that were used in the Concorde supersonic aircraft, particularly information regarding the temperatures that these rivets would attain and their properties when the skin is subjected to skin friction heating. Therefore, the effect that various temperatures would have on the rivet's strength, and, in turn, whether the rivets will be durable to maintain attachment between the skin and the internal structural members (the wing is one very principal example of such an attachment), are imperative to understand. It therefore is practical to use titanium alloy rivets in supersonic aircraft from the discussions above, and beneficial to tuck the rivet heads flush or countersunk with the skin of the aircraft. The excessive speed of the Concorde or any supersonic or hypersonic aircraft, space shuttle, rocket, military projectile, and so on, would subject any protruding surface to severe airflow causing either damage or excessive heating, which would ultimately weaken the material and possibly leading to eventual failure. Also objects such as hail, birds, or even rain droplets when impacting directly upon the exposed and protruding rivet heads in aircraft, for instance, would almost certainly cause physical damage. The effects of water droplets on a supersonic aircraft are most likely to be less severe than other materials that may impact the aircraft. Water is highly likely to disintegrate and turn into mist due to aerodynamic airflow, shock waves, and the surface temperature of the aircraft. The effects of airflow will affect lift and drag depending upon surface roughness or items protruding from the aircraft skin surface; however, these in turn will also be affected by increased friction and the increased skin surface temperature of the aircraft. The skin which is connected to the rivets generates high temperatures during supersonic flight, but it would be difficult to place samples of the small rivets to be tensile tested by conventional means. Certain regions of the rivets might have been exposed to higher temperatures, and ideally either their mechanical properties or remaining life span may be determined if their mechanical properties and their corresponding microstructural evolution are better understood. Since the majority of rivets are relatively small components, performing tensile tests on rivets in many cases may not be feasible. One method of extrapolating a rivet's tensile strength would be to perform hardness tests on the rivets and then convert these hardness tests into tensile strength values.

3.2 DELIBERATIONS ON GENERATING TEMPERATURE AND HARDNESS DATA FROM RIVETS IN SUPERSONIC FLIGHT

Our research is continually being developed at the University of New South Wales, Sydney. As a result, to further understand the effect of aerodynamic heating due to high Mach numbers on aerospace rivets, further analysis should be conducted using the ANSYS Fluent software. Alternatively, analysis could be conducted on SolidWorks Flow Visualization, as both of these programs allow for Computational Fluid Dynamics (CFD) modelling, and they also provide analysis to accurately predict the temperature at any point on the surface of the aircraft, including the average temperature of the rivet. Some points to consider for future qualitative experimental research in this field should include the following:

1. It is difficult to understand the microstructural properties of a rivet that have been simulated on ANSYS Fluent or SolidWorks Flow Visualization since only the temperature of the surface of the rivet can be ascertained. The speed range of supersonic wind tunnel testing ranges between Mach 1.2–5, and wind tunnel testing must be conducted in order to simulate and achieve an understanding of any microstructural changes within the rivet as a result of aerodynamic heating. Therefore, procurement of a titanium-based alloy similar to the ones used in the Concorde and subjecting it to testing in a wind tunnel (for at least 30 minutes, and up to a maximum of 60 minutes so that the effect of aerodynamic heating is achieved on the rivet) is recommended.

2. Examination of the microstructure of the rivet using a transmission electron microscope (TEM) before and after wind tunnel testing to verify whether aerodynamic heating has contributed any appreciable microstructural changes in the vicinity of either the rivet's surface or in its inner regions. Assessing microstructural changes that would be evident in the rivet (as mentioned above in point 1), due to aerodynamic heating within the rivet. Typical aerospace aluminium alloy 7075 and 5056 rivets were used to prepare samples for TEM analysis by Melhem et al. at the School of Materials Science and Engineering, University of New South Wales, Sydney (UNSW) laboratory, and the results were published in the paper titled "Field Trials of Aerospace Fasteners in Mechanical and Structural Applications" [54]. The samples were prepared for TEM analysis focused ion beam microscopy using an FEI Nanolab 200. Platinum was used to protect the surface of the specimens in situ prior to applying the ion beam, and subjected to an FEI Tecnai GS at an accelerating voltage of 200 kV. However, for the sake of the rivets used in the Concorde, the alloy need not be of an aluminium alloy composition for the skin connection to the frame as it would not have been able to sustain the loading and creep resistance requirements, rather it would have been manufactured from titanium alloy.

3. Wind tunnel testing is required that simulates the Concorde's conditions; this includes the Mach number and the atmospheric temperature. To obtain adequate aerodynamic heating on a rivet, the wind tunnel testing duration should be between 30–60 minutes (as mentioned above in point 1).

4. Examine the microstructure of the rivet during the various stages of wind tunnel testing for both the heating and cooling cycles using a Confocal

Scanning Laser Microscopy (CSLM). In turn, an evaluation can be made about the specific microstructural changes that develop from the surface of the rivet to the centre of the rivet. This method does not require the rivets to be cut into thin sections, and the real-time thermodynamic calculations can be performed based upon observing in situ metallurgical phase transformations. Although the temperatures on the Concorde surface would at most reach 128°C, this temperature would need to be adhered to and controlled in wind tunnel testing so that the rivet simulates the conditions during flight. It would also be useful to try and simulate a small part of the construction of the aircraft with the rivet and skin attached through to a comparable internal structure that is used in aircraft, and ensure allowance for interference fits in rivet holes which would also be identical to the real aircraft construction. Generally, rivets in subsonic aircraft that are used to connect internal structures to the skin require a transition or interference fit as the aircraft is subjected to various loading while in-flight or even on the ground. Ultimately, the interference fit allows for the material around the rivet hole to be compacted so that it permits excellent load transfer. The structure would have to be enclosed so that the true nature of any aerodynamic heating effects generated from the skin and on the rivet in the actual aircraft would be replicated in wind tunnel testing. It has been suggested by Lundberg [88] that the shank of the rivet does not undergo any heating, as airflow does not extend down the rivet shank or inner spars because the skin acts as a barrier.

5. Examination of the microstructure of the rivet by visual inspection, optical microscopy, and Vickers microhardness testing. A correlation can then be made between the microstructure observed and the Vickers microhardness or Vickers hardness number (VHN) to determine the ultimate tensile strength (UTS) and yield strength (YS) due to residual frictional heating that could have possibly been generated from the skin to the rivets. Typical samples prepared for Vickers microhardness testing by the authors Melhem et al. [54], and the samples were prepared in the School of Materials Science and Engineering, University of New South Wales, Sydney (UNSW) laboratory as follows:

• Rivets were prepared and sectioned with a Struers Minitom diamond saw using cooling oil. The sections were cold mounted and polished. Cold mounting was used rather than hot mounting, as hot mounting will add extra heat to the rivet section and therefore would potentially subject the aluminium alloy rivet sections and their grain boundaries to further ageing kinetics. The resin holding the rivet sections were prepared in a Buehler Cold Mounting unit, using one part hardener to four parts epoxy. This resin solution was placed inside a vacuum pump for five minutes in order that any excess air bubbles are extracted. Vaseline was applied to the surface so that it would be lubricated for ease of removal. Silicon carbide polishing pads were used with grit levels ranging between 120, 320, 1200, and 2500. The grinding of the sample surface was performed until it was coplanar due to applying low speeds of 200 rpm on a Struers Labopol 5.

• The samples were cleaned prior to polishing by applying cotton wool under running water and afterwards immersing it in soapy water in an ultrasonic bath for two minutes. Ethanol was applied to the surface to dry the sample, along with compressed air blow-drying. Further polishing was applied to the sample by using the Struers Labopol 5 machine at 150 rpm, using a 3 micron (µm) oil-based polycrystalline Kemet Diamond suspension on a Kemet NSH-BH polishing pad, followed by 1 µm using 1/10 µm oil-based polycrystalline Kemet Diamond suspension. The final steps involved polishing with oxide polishing-colloidal silica suspension and a Kemet Chem-HM polishing pad. An Olympus PMG 3 microscope at a magnification of 25x was used to inspect the polished surface, prior to etching the sample on a Nikon Epiphot 200 microscope at 100x magnification. Keller's Reagent (1.5 ml HCl, 1.0 ml HF, 2.5 ml HNO_3, 95 ml distilled water) was used to etch the samples and this was performed under a fume hood at room temperature. The samples were immersed in the solution for 30 seconds prior to being cleaned with ethanol.

Figure 3.1 shows the aerospace rivet that was used for the research by the authors Melhem et al. [54] as mentioned above for the TEM and Vickers hardness examination, but this rivet was of aluminium alloy and was used for Boeing subsonic aircraft. This is only an illustration to demonstrate typically the cross sections of rivets and the types of testing and assessments that can be performed similarly on a titanium alloy rivet that would be used on a supersonic aircraft. **Figure 3.2** shows a cross section of an installed rivet and some of the characteristics deriving from its design.

1 = Flush installation makes fastener easier to paint and resistant to salt and water.
2 = Flush pin break eliminates need for grinding/filing.
3 = Solid-circle lock ensures maximal strength and resistance to vibration - designed to resist pin pushout.
4 = Sleeve expands during installation, tightly filling hole to create a weather-resistant joint.

Figures 3.3–3.5 show a schematic of the samples cross section that have been prepared for the mounting process. The designation numbers used were 1, 2, and 3, where the first number refers to the sample (1, 2, or 3) and the second number refers to location of the analysis.

Vickers microhardness measurements were performed using a Stuers Duroscan G5 with a weight of 300 g applied for a dwell time of 15 seconds.

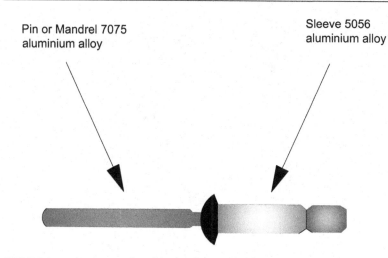

FIGURE 3.1   Aerospace Rivet Used in Present Work [54].

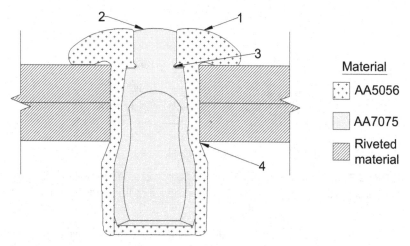

FIGURE 3.2   Cross Section of Installed Rivet and Characteristics [54].

Indent images were recorded using a Nikon Epiphot 200 microscope. The data was tabulated with the location, the Vickers microhardness, and the alloy type for comparative purposes.

An extrapolation of hardness to tensile strength for rivets would allow for quicker inference of durability in aircraft. However, the below mention of ageing is only relevant to aluminium alloys and not to titanium alloys. Nonetheless, an appreciation can be made by this scenario, since a metal's hardness is not a true representation of its mechanical properties such as

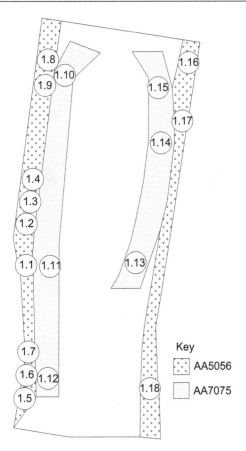

FIGURE 3.3 Sample 1 – Longitudinal Cross Section of 12-Year-Old Rivet, Showing Mandrel on Inside, Sleeve on Outside, and Locations for Microhardness Testing [54].

tensile strength, since at similar hardness values some metals have higher tensile strength values due to differing heat treatment. The strength of an alloy is therefore more dependent upon the alloy's evolvement of their microstructure rather than the alloy's hardness value.

For instance, the measure of wear resistance (inverse of abrasive wear or loss of material) and bulk hardness has been established to exist as a linear relationship for pure metals [89, 90]. On the other hand, the measure of wear resistance of pure metals was found to be lower when compared to that of heat-treated steels with the same hardness [89]. In addition to this, it was found that a single steel heat treated to various levels of hardness did not exhibit a linear relationship between wear resistance and hardness [90]. This suggests that the microstructure is what determines the wear resistance of materials, and

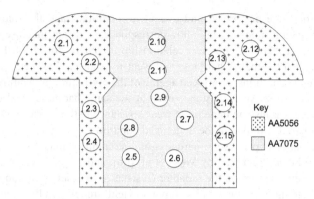

FIGURE 3.4 Sample 2 – Longitudinal Cross Section of 12-Year-Old Rivet, Showing Mandrel on Inside, Sleeve on Outside, and Locations for Microhardness Testing [54].

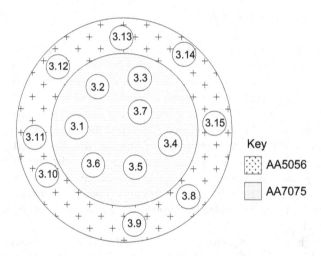

FIGURE 3.5 Sample 3 – Diametral Cross Section of New Rivet, Showing Mandrel on Inside, Sleeve on Outside, and Locations for Microhardness Testing [54].

Moore [91] found in his research that the wear of ferritic steels was not only dependent upon the bulk hardness but was also dependent upon microstructure. Research on an aluminium alloy microstructure, for instance in 2124 aluminium alloy, shows that the highest wear resistance was possible only when it reached peak hardness when comparing to materials that have been underaged and of the same hardness [92, 93]. The load indenter, which has different loads when taking hardness measurements with the Vickers microhardness, is complex and difficult to understand [94]. Melhem et al. [95, 96] researched the relationship between the wear resistance and hardness of 6061 aluminium alloys in both the underaged and overaged conditions, and the findings for increased ageing times of this alloy corresponded with an increase in wear resistance, but there was no clear relationship in general for all of the alloys tested between the matrix hardness and wear resistance. However, within the experimental conditions, prolonged ageing of the 6061 aluminium alloys, there was an increase in wear resistance even at the same hardness level of these alloys. There was also an increase in wear resistance in the overaged condition when compared with that in the underaged condition.

The relationship between hardness and strength of materials of various kinds have been researched by Voort et al. [98], and they show some correlation between pressure on the indenter (resistance to indentation) and that of the tensile strength, yet the relationship with yield strength is a little more complex. Voort has related yield strength for carbon steels to the Vickers hardness as follows:

$$\text{YS (kgf/mm}^2) = 1/3 \text{ VHN } (0.1)^{m-2}$$

where m is the Meyer's strain hardening coefficient [97] and kilogram-force is *kgf*. This relationship only applies to carbon steels since the association between hardness and yield strength is dependent upon the strengthening mechanism. Aluminium alloy steels, for instance, exhibit an increased strain hardening coefficient when they have been aged and when compared to carbon steels, but they have lower yield strengths than cold worked alloys [98]. Although there have been findings that relate hardness and tensile strengths in certain steels and alloys, there is still a lack of understanding as these relationships are only empirically based, and more experimentation is required to confirm a precise relationship between hardness and strength for any particular material. Further information relating to mechanical properties and hardness can be found in Datsko et al. [99].

Through empirical equations, it appears that the VHN can be converted to YS and UTS. Pagliarello [100] has provided equations that were created from data and equations were also adopted by Clark et al. [101] for altered 7075-T6 specimens. *Equations (3.1)–(3.5)* below are empirical equations, where *Equation (3.1)* relates the conversion of VHN to Hardness-Rockwell B (HRB). These methods require the Vickers microhardness to be converted to the Rockwell B hardness, which uses the data of ISO 18265:2013 Metallic Materials – Conversion of Hardness Values. *Equations (3.2)* and *(3.3)* provide the calculation for YS and UTS respectively by Pagliarello [100], and *Equations (3.4)* and *(3.5)* provide the calculation for YS and UTS respectively by Clark et al. [101]:

Equation (3.1): HRB = 0.3 x H_v + 34.987 (140 ≤ Hv ≤ 195), using the data of *ISO 18265:2013 Metallic Materials – Conversion of Hardness Values*);

Equation (3.2): YS = 9.75 x HRB - 388.19   [100],

Equation (3.3): UTS = 9.49 x HRB - 273.64 [100],

Equation (3.4): YS = 8.92 x HRB - 309.95   [101],

Equation (3.5): UTS = 6.11 x HRB + 0.76   [101].

The results for the conversions showed that the data for YS agreed relatively well (1–2% difference), while the data for UTS were not in agreement (4–7% difference). Hardness testing of a titanium alloy rivet can be conducted in a similar method as described above for the 6061 aluminium alloy. It would be beneficial to reiterate the properties of titanium that were mentioned earlier, in order to determine which alloy is most likely and more functional for use in rivets in supersonic aircraft. The following provides the typical characteristics properties of the common titanium alloys: alpha-titanium (α-Ti) is not heat treatable, it can be easily welded, it has excellent creep resistance and ductility at high temperatures, and it is used mainly in engines requiring high temperature service. Beta-titanium (β-Ti) is heat treatable and can also be welded. It has high tensile strength and fatigue resistance, and is mainly used in the SR-71 Blackbird, fuselage, wing, body skins, longerons, and ribs. Alpha + beta-titanium (α + β-Ti) are heat treatable and are used in jet engines and in airframes due to the combined properties of α-Ti & β-Ti; however, the high strength creep in this alloy is not as good as in the α-Ti. Titanium (Ti-6Al-4V) is the "workhorse" of titanium and forms the α+ β-Ti group. It is heat treatable and used for high load requirements such as in aerospace fasteners, turbine engines, skins, wing box, stiffeners, and spars.

Generally, if titanium alloys are not heat treatable, but are weldable and maintain properties such as creep resistance and ductility, then they would be useful at high temperatures such as in the engine. If titanium alloys are both heat treatable and weldable, and yet maintain high strength and resistance to fatigue, then they would be useful in the majority of the structural parts of a military or supersonic aircraft. The most commonly sought after alloy in the titanium family of alloys is Ti-6Al-4V, since they are heat treatable and are applied in areas requiring high loading and strength. Generally, Ti-6Al-4V alloys are used as rivets, fasteners, and bolts in aircraft since they have high strength, good corrosion resistance, heat resistance, and creep resistance [102]. Titanium rivets can be depended upon to maintain structural stability such as hole and shank tolerances, which are both particularly important for rivets and fasteners. The yield strengths in these titanium alloys range between 760–1,300 MPa on average, and the tensile strengths range between 790–1,580 MPa. It is therefore imperative to place in perspective the skin temperature of the Concorde which travelled at Mach 2.0, as the temperature reached was insignificant (maximum 128°C on nose of Concorde) when comparing it to the temperatures of, for instance, the engines that need to combat high temperatures. The Concorde's skin generally was not exposed to much higher temperatures than 128°C, thereby the design allowed for moderate thermal expansion in various parts of the aircraft, such that corrugated structural webs and slits in the leading edges were incorporated, and it is worth recalling that the entire Concorde fuselage did expand by about 300 mm overall while in full supersonic flight. However, titanium in contact with metals such as aluminium that are far apart in the galvanic series may cause galvanic corrosion if precautions to alleviate these issues are not taken by applying certain coatings such as zinc chromate metal spray primer before and after installation. Overall, when comparing titanium rivets to aluminium rivets, titanium has higher strength than aluminium, good corrosion resistance, and can be used at higher temperatures since it has far superior fatigue life endurance. Titanium alloy is a better choice of material for supersonic and hypersonic aircraft application.

REFERENCES

54   Melhem GN, Munroe PR, Sorrell CC, et al. Field Trials of Aerospace Fasteners in Mechanical and Structural Applications. In: Totten GE, Tiryakioglu M and Kessler O (eds), *Encyclopedia of Aluminium and Its Alloys*. Boca Raton, FL: CRC Press, 2019, pp. 1007–1021.

84   Nygren WD. Development of a High Force Thermal Latch. *29th Aerospace Mechanisms Symposium hosted at South Shore Harbour Conference Facility, League City, TX*, 1995, p. 164.

88   Peel, C.J. Degradation of Aluminum Alloys—Concorde Experience. Paper presented at the Workshop on Long-Term Aging of Materials and Structures, National Materials Advisory Board, National Research Council, Washington, D.C., August 10–12, 1994.

89   Kruschov MM and Babichev MA. *Resistance to Abrasive Wear of Structurally Inhomogeneous Materials, Friction and Wear in Machinery*. New York City: ASME, 1958, p. 5.

90   Mutton PJ and Watson JD. Some Effects of Microstructure on the Abrasion Resistance of Metals. *Wear*, 48, 1977, pp. 385–398.

91   Moore MA. The Relationship Between the Abrasive Wear Resistance, Hardness and Microstructure of Ferritic Materials. *Wear*, 28, 1974, pp. 59–68.

92   Wang A and Rack HJ. *Metal and Ceramic Matrix Composites: Processing Modelling and Mechanical Behaviour*. Warrendale, PA: The Mineral, Metals and Materials Society, 1990, p. 487.

93 Hutchings IM. *Abrasive and Erosive Wear of Metal-Matrix Composites.* 2nd European Conference on Advanced Materials and Processes, Cambridge, 1991. (Forthcoming in the Institute of Metals, London, proceedings Euromat '91.)

94 Manson W, Johnson PF, and Varner JR. Importance of Load Cell Sensitivity in Determination of the Load Dependence of Hardness in Recording Microhardness Tests. *Journal of Materials Science*, 26, 1991, pp. 6576–6580.

95 Melhem G, Bandyopadhyay S, and Krauklis P. The Influence of Artificial Ageing on Abrasive Wear of SiC and Al2O3 Particulate Metal Matrix Composites. *Proceedings of the 3rd Australian Forum on Metal Matrix Composites, at the School of Materials Science and Engineering, University of New South Wales Kensington, NSW, 2033, Australia*, 1992, pp. 45–62.

96 Melhem G, Krauklis P, Mouritz AP, et al. Abrasive Wear of Thermally Aged 6061 Al-SiC Composites. *Proceedings of the Third International Conference on Composites Engineering (ICCE/3)*. New Orleans, LA, 1996, pp. 577–578.

97 Fee A. Selection and Industrial Application of Hardness Tests. *ASM Handbook, Vol. 8: Mechanical Testing and Evaluation*, Edited by Kuhn H and Medlin D, Materials Park, OH: ASM International, 2000, pp. 260–277.

98 Voort GFV. Hardness. In: George Voort (ed), *Metallography: Principles and Practices*. Materials Park, OH: ASM International, 1999, pp. 383–385, 391–393.

99 Datsko J, Hartwig L, and McClory B. On the Tensile Strength and Hardness Relation for Metals. *Journal of Materials Engineering and Performance*, 10(6), 2001, pp. 718–722.

100 Pagliarello AG. *Effects of Modified Solution Heat Treatment on the Mechanical Properties and Stress Corrosion Cracking Susceptibility of Aluminum Alloy 7075.* PhD thesis, Carleton University, Ottawa, Ontario, 2011.

101 Clark R, Coughran B, Traina I, et al. On the Correlation of Mechanical and Physical Properties of 7075-T6 Al Alloy. *Engineering Failure Analysis*, 12(4), 2005, pp. 520–526.

102 Welsch G, Boyer R, and Collings EW. *Materials Properties Handbook: Titanium Alloys*. Almere, The Netherlands: ASM International, 1993.

Conclusion

In this review, we have presented a brief finding of the literature behind super-sonic aircraft, endeavouring to emulate what is known about the Concorde as the basis of our investigation. The greatest incentive that the Concorde afforded ardent travellers in the past was a substitute for subsonic to super-sonic velocity from one destination to another, thereby reducing their time by about half – essentially, *passengers were in a time machine.* The multitude of experts in the engineering and science disciplines still need to work together in designing a supersonic passenger aircraft to ensure a safer, environmentally friendlier and more economical aircraft for the use of passenger flights than the Concorde. Although the Concorde can provide more than just a stepping stone for future consideration of developing supersonic aircraft for everyday passengers, historical data through microscopic analysis for most materials on supersonic aircraft including the Concorde have not been well characterised nor well-documented, leading to a "gap" in knowledge. This could be due to one of two principal reasons: either aerospace designers/manufacturers must closely guard their secrets against their competitors, or data in the field is still in its infancy stage and not very well understood. It is clear from the exten-sive literature review that the investigators assessing the wreckage and trying to ascertain information relating to materials in the Concorde that crashed in 2000 found it exceedingly difficult to do so. The Concorde was designed and constructed to travel at Mach 2.0, but the materials, technology, and envir-onmental impact required to maintain this speed was different to what was conventionally available for subsonic and military aircraft. In the above, some points have been presented for future consideration into qualitative experi-mental research, which would ideally shed further light on certain materials subjected to supersonic effects.

A preliminary assessment in this review has been made for titanium rivets, which could be tested under conditions that simulate aerodynamic effects when the aircraft encounters speeds in excess of Mach 2.0. Furthermore, it is anticipated from future research that the data from the rivets would be generated in the laboratory. For instance, wind tunnel experiments would be the basis for the aerodynamic investigation on titanium rivets, and from these experiments, an analysis should be carried out by sectioning the rivets and performing detailed microstructural evaluations. This would also help to determine whether a rivet's strength varies with increasing time and increasing temperatures during flight at supersonic speed, therefore, Vickers microhardness evaluation is also required so that if the rivet has aged

DOI:10.1201/9781003516903-4

further due to extra heating and time, then a correlation could be made with overageing of the rivets. Certain empirical equations and methods of hardness testing on rivet sections have been presented in Chapter 3 in order to convert VHN to YS and UTS. Whilst research in literature shows that ageing of certain aluminium alloys leads to increased tensile strength, there is a limit whereby the ageing time at a constant temperature or varied temperatures will affect the ageing kinetics and in turn the strength of the alloy. The skin of the Concorde is made from the Hiduminium aluminium alloy, but the rivets are made from titanium alloy. Although these are different materials, it is clear that the temperatures and duration during supersonic flight for both aluminium alloys and titanium alloys are not identical, nor are their mechanical properties. The properties of titanium are far superior to that of aluminium alloys and this is the reason aluminium alloys are restricted in the use of aircraft. However, the variations in temperature, and pressurisation/depressurisation of the cabin, stresses on the wings, fuselage et cetera all affect the performance of the titanium alloy rivets that hold many items within these parts of the aeroplane in the long run.

Amongst the myriad of materials used in the Concorde, meticulous design and careful quality control for the construction of the aircraft would have ensured that materials far apart in the galvanic series are separated, or at the very least are insulated by some suitable protective means from one another. Also, knowledge of the CTE of the respective materials is imperative, as large differences would be detrimental to the structures as they potentially create stresses, leading to eventual failure. The use of the RR58 aluminium alloy skin would have been in contact with several other metals in the Concorde. For instance, titanium alloy rivets or fasteners are in contact with RR58, and are irrespective of any protective measures that would have been taken to insulate the materials from one another; these materials combined would be subjected to a large difference in electrical potential (materials that are far apart in the galvanic series), thus placing these metals at high risk of corrosion. Therefore, if the rivets or fasteners are manufactured from titanium alloy (cathodic), and the skin is manufactured from aluminium alloy (anodic), then the ratio of anode to cathode is large, thereby reducing, not eliminating, the likelihood of corrosion. Also, when Ti alloys and aluminium alloys are exposed to temperatures between 20–200°C which was typical in the Concorde, then these alloys are predisposed to excessive thermal expansion differences (aluminium alloy 2618-T6 expands almost 2½ times more than the titanium alloy Ti-6Al-4V alloy). Creep strain would also be experienced by such thermal expansion and stresses causing hairline cracks, and these cracks lead to fatigue cracks which ultimately causes failure of that section. It should be noted that small cracks were found in all seven of the Concorde wings which were reported by British Airways only one day prior to the Concorde crash by Air France. However, British Airways quashed the findings of these cracks as being of no major concern or consequence. There are no records or information in any literature that provide reasons as to why or how these cracks were caused. However, when the Air France Concorde crashed due to a titanium strip lying on the runway in France, by default, any other major metallurgical issues that were present in the Concorde aircraft also vanished with the entire Concorde aircraft when they were permanently retired in 2003, three years after the Concorde crash.

The intention to systematically evaluate rivets experimentally as discussed in this book would be primarily to both improve the understanding of aerodynamic heating on the rivets and to also gain a better understanding of the effects of adjoining materials and structures due to their connection with the rivets. By improving the understanding of the material's microstructural processes under supersonic conditions, it should be possible to understand further whether newer materials and processes can be developed, in order to improve upon the limitations that have been experienced to date. Ultimately, this research should provide more than just an advancement for further investigations on materials subjected to the effects of supersonic flight, and it should generate further attention and interest in order to narrow the "gap" of knowledge in this field.

Index

Printed in the United States
by Baker & Taylor Publisher Services